Sea-Dumped Chemical Weapons: Aspects, Problems and Solutions

NATO ASI Series

Advanced Science Institutes Series

A Series presenting the results of activities sponsored by the NATO Science Committee, which aims at the dissemination of advanced scientific and technological knowledge, with a view to strengthening links between scientific communities.

The Series is published by an international board of publishers in conjunction with the NATO Scientific Affairs Division

A Life Sciences **B Physics**	Plenum Publishing Corporation London and New York
C Mathematical and Physical Sciences **D Behavioural and Social Sciences** **E Applied Sciences**	Kluwer Academic Publishers Dordrecht, Boston and London
F Computer and Systems Sciences **G Ecological Sciences** **H Cell Biology** **I Global Environmental Change**	Springer-Verlag Berlin, Heidelberg, New York, London, Paris and Tokyo

PARTNERSHIP SUB-SERIES

1. Disarmament Technologies	Kluwer Academic Publishers
2. Environment	Springer-Verlag / Kluwer Academic Publishers
3. High Technology	Kluwer Academic Publishers
4. Science and Technology Policy	Kluwer Academic Publishers
5. Computer Networking	Kluwer Academic Publishers

The Partnership Sub-Series incorporates activities undertaken in collaboration with NATO's Cooperation Partners, the countries of the CIS and Central and Eastern Europe, in Priority Areas of concern to those countries.

NATO-PCO-DATA BASE

The electronic index to the NATO ASI Series provides full bibliographical references (with keywords and/or abstracts) to more than 50000 contributions from international scientists published in all sections of the NATO ASI Series.
Access to the NATO-PCO-DATA BASE is possible in two ways:

– via online FILE 128 (NATO-PCO-DATA BASE) hosted by ESRIN, Via Galileo Galilei, I-00044 Frascati, Italy.

– via CD-ROM "NATO-PCO-DATA BASE" with user-friendly retrieval software in English, French and German (© WTV GmbH and DATAWARE Technologies Inc. 1989).

The CD-ROM can be ordered through any member of the Board of Publishers or through NATO-PCO, Overijse, Belgium.

Series 1: Disarmament Technologies – Vol. 7

Sea-Dumped Chemical Weapons: Aspects, Problems and Solutions

edited by

Alexander V. Kaffka
Institute of the USA and Canadian Studies,
Russian Academy of Sciences,
Moscow, Russia

Kluwer Academic Publishers

Dordrecht / Boston / London

Published in cooperation with NATO Scientific Affairs Division

Proceedings of the NATO Advanced Research Workshop on
Sea-Dumped Chemical Munitions
Kaliningrad (Moscow Region), Russia
January 12–15, 1995

A C.I.P. Catalogue record for this book is available from the Library of Congress

ISBN 0-7923-4090-6

Published by Kluwer Academic Publishers,
P.O. Box 17, 3300 AA Dordrecht, The Netherlands.

Kluwer Academic Publishers incorporates the publishing programmes of
D. Reidel, Martinus Nijhoff, Dr W. Junk and MTP Press.

Sold and distributed in the U.S.A. and Canada
by Kluwer Academic Publishers,
101 Philip Drive, Norwell, MA 02061, U.S.A.

In all other countries, sold and distributed
by Kluwer Academic Publishers Group,
P.O. Box 322, 3300 AH Dordrecht, The Netherlands.

Printed on acid-free paper

Printed in the Netherlands

CONTENTS

Preface

This volume summarises the materials presented at the NATO Advanced Research Workshop on Sea-Dumped Chemical Munitions, held in Kaliningrad (Moscow Region), Russia, in January 1995. The conference was sponsored by the NATO Division of Scientific and Environmental Affairs in the framework of its outreach programme to develop co-operation between NATO member countries and the Cooperation Partner countries in the area of disarmament technologies.

The problem of the ecological threat posed by chemical weapons (CW) dumped in the seas after the Second World War deserves considerable international attention: the amount of these weapons, many of them having been captured from the German Army, is assessed at more than three times as much as the total chemical arsenals reported by the United States and Russia. They were disposed of in the shallow depths of North European seas - areas of active fishing - in close proximity to densely populated coastlines, with no consideration of the long-term consequences. The highly toxic material have time and again showed up, for instance when retrieved occasionally in the fishing nets, attracting local media coverage only.

Nevertheless, this issue has not yet been given adequate and comprehensive scientific analysis, the sea-disposed munitions are not covered by either the Chemical Weapons Convention or other arms control treaties. In fact, the problem has been neglected for a long time on the international level. Only recently were official data made available by the countries which admitted conducting dumping operations.

There were a number of reasons for the decades of delay in addressing this problem, during which time the containers and shells loaded with combat CW were deteriorating in the sea water. The government bodies of both the states that carried out the dumping operations and those bordering the dumping areas were reluctant to tackle this sensitive problem, especially during the period of East-West tension. With the Cold War now over, the political obstacles to addressing this problem have mostly been removed. However, there remains the extreme scientific and technical complexity of the problem posed by the CW dumps, which requires comprehensive and profound expertise.

The conference in Kaliningrad was the first attempt to address this issue on a comprehensive scientific basis. It was organised by the NATO Division of Scientific and Environmental Affairs and the "Conversion For The Environment" Foundation (CFE). CFE, an international non-governmental organisation conceived to address the ecological problems caused by military activities and the environmental perspectives of economic conversion, concentrated on the problem of sea-dumped CW and saw its mission in drawing due attention to this previously neglected problem. The conference project started when the Division of Scientific and Environmental Affairs showed interest in the theme of our proposal for an Advanced Research Workshop in 1992, well before the subject of sea-dumped CW was put on the Europe's official agenda in the framework of the *Ad Hoc* Working Group established by the Baltic Marine Environment Protection Commission.

The conference's concept envisaged consideration of the various aspects of the problem: chemical, biological, technological. The conference demonstrated the availability, in different countries, of substantial relevant knowledge, experience, technologies and other resources that can be applied to the particular ecological problem of sea-dumped CW. It also showed the lack of awareness of each other's achievements and capabilities, and confirmed the need for improved information exchange and co-ordination. The conference, which drew together experts from the centres of academic and applied research from Europe and the United States, approved a concrete set of recommendations based on the current level of scientific knowledge on the problem.

An important feature of the conference's concept was the involvement of representatives from defence-related industries and research centres: the economic conversion of the defence sector in the East and West offers new opportunities to apply this sector's vast experience and resources in areas of environmental concern. The ecological problem of sea-dumped CW in particular not only has direct relevance to the military industry, but requires much specific experience (i.e. in weapons destruction or underwater technologies) which may be obtained only from the defence industry and science of the countries concerned.

The Conference on sea-dumped chemical munitions was made possible due to a grant from NATO. It can be seen as another sign of modern times, when defence organisations, within the framework of new partnerships, contribute to solving the common ecological threats Europe is facing today, and help re-pay the environmental debt caused by the military. The conference can be seen as a positive effort toward putting the environment and defence industries in concert.

As mentioned above, the issue of sea-dumped CW is a particularly multi-faceted problem. Besides it essential chemical and biological components, it incorporates important historical and legal implications. It also represents a complicated technological challenge, and requires economic assessment as well. An exhaustive analysis of all these aspects of the issue is beyond the framework of a single conference, and we hope they will be properly addressed in the programmes to come. However, one of the achievements of the unprecedented conference in Kaliningrad was that it was able to raise virtually all these questions in its agenda and lay a foundation for further comprehensive applied research.

While some of the papers presented at the conference were of an interdisciplinary nature, touching on, *inter alia,* historical or legal questions, most presentations had either a chemical or a technological focus, and the book chapters were organised accordingly. However, the attribution of the chapters to either "chemical" or "technological" divisions is not strict. It should also be noted that, given the close interrelation between the problems of chemical weapons disposed in seas and of those stored on land, some of the papers provide substantial coverage of the problems associated with the storage, destruction, and environmental effects of on-land CW as well.

Alexander Kaffka
Workshop Director

Acknowledgements

The production of this volume was a collaborative process, and there are many to whom the editor is appreciative. The book would not exist were it not for NATO's sponsorship and invaluable co-operation, which included critically important consultations with Prof. H. Schubert, then director of the Fraunhofer Institute of Chemical Technology (Germany), as well as other assistance on numerous occasions. I am thankful to the respected contributors for their high-quality and informative articles. Their responsiveness to the editor's requests and the ability to produce updated versions of their pieces in the required format within a tight production schedule is indeed admirable. I would also like to thank Boston University's Division of International Programs (Moscow) and Ross Colgate for assistance in editing the translated texts.

Conference On Sea-Dumped Chemical Munitions
Participants List

International participants:
Dr. E. Andrulewicz (Poland) - Institute of Meteorology and Water Resources, Gdyna;
Dr. Mark Frondorf (USA) - Science Applications International Corporation, McLean, VA;
Mr. Howard Graeffe (USA) - National Institute for Environmental Renewal (NIER), Mayfield (PA);
Dr. Per Olof Granbom (Sweden) - FOA NBC Defense Research Establishment, Umea;
Dr. H. Haeber (Germany) - Heinrich Hirdes GMBH, Berlin;
Dr. Joerg Hedtmann (Great Britain) - SubSea Offshore Ltd, Aberdeen;
Dr. Jim Henry (USA) - University of Tennessee, Chattanooga;
Mr. Iain Jarvies (Great Britain) - SubSea Offshore Ltd, Aberdeen;
Mr. Randy Kritkausky (USA) - Director, "ECOLOGIA", La Plume, PA;
Dr. Anthony P. Malinauskas (USA) - Martin Marietta Energy Systems, Oak Ridge, TN;
Mr. Stefan Robinson (Switzerland) - Chemical Weapon Program, Execitive Committee, The Green Cross International;
Ms. Nancy Schulte (Belgium-NATO) - Program Director, Scientific and Environmental Affairs Division, NATO HQ;
Dr. Thomas Stock (Sweden) - SIPRI (Stockholm International Peace Research Institute), Chemical and Biological Weapons Program, Solna, Sweden;
Mr. Lynn Taylor (USA) - PAR Government Systems, Arlington, VA;
Dr. F. Volk (Germany) - Fraunhofer Institute for Chemical Technology, Pfinztal Berghausen;
Dr. G.F.White (United Kingdom) - Department of Biochemistry, University of Wales College of Cardiff, Cardiff;
Prof. Zygfryd Witkiewicz (Poland) - Institute of Chemistry, Military University of Technology;

Russian participants:
Prof. Kirill Babievsky (Russia) - Nesmeyanov Institute of organic compounds, Russian Academy of Sciences, Moscow;
Prof. Sergey Baranovsky (Russia)- Vice-President, The Green Cross Russia;
Dr. Tengiz Borisov (Russia) - Ministry of Defense;
Igor Dobrov (Russia) - Karpov Institute of Physical and Chemical Research, Obninsk branch
Igor Epifanov (Russia) - Corporation "Mashinistroenie";
Prof. Lev Fyodorov (Russia) - Chairman, Union for Chemical Security, Vernadsky Institute of Geochemistry and Analitical Chemistry ;

Dr. Lyudmila Kalinina (Russia) - Institute of General Genetics, Russian Academy of Science, Moscow;
Dr. Vyacheslav Konykov (Russia) - President, Conversion Technologies Corporation, Moscow District;
Mr. Kuznetsov O.L. - (Russia) Ministry of Emergencies and Elimination of Consequences of natural Disasters
Gen.-leut. (ret.) Yuri Lebedev (Russia) - Russian-American University;
Prof. Georgy Lisichkin (Russia) - Head of Laboratory, Chemical Faculty, Moscow State University;
Captain (Ret.), Dr. Leonid Malyshev (Russia) - Ministry of Emergencies and Elimination of Consequences of Natural Disasters;
Dr. Andrey Melvil - Deputy Chairman, Russian Federation for Peace and Reconciliation
Academician, Dr. Vladimir Morozov (Russia) - Chief Designer, Vympel Joint-Stock Corporation, Moscow;
Dr. Boris Nersesov (Russia) - Head of section, Russian Academy of Natural Sciences;
Prof. Sergey Novikov (Russia) - Head of Department, Semyonov Institute of Chemistry;
Dr. Victor Plotnikov - (Russia) Karpov Institute of Physical and Chemical Research, Obninsk branch;
General-Major (Ret.) Dr. Boris T. Surikov (Russia) - Institute of the USA and Canada, Russian Academy of Sciences, Moscow;
Dr. Olga Stemberg (Russia) - Russian Ministry of Environment;
Dr. Valery Sevostyanov (Russia) - member of the Moscow City Duma (parliament);
Mr. Igor Vlasov (Russia) - Presidential Committee on CBW conventional problems;
Professor Sergey Yufit (Russia) - Institute of the Organic Chemistry, Russian Academy of Sciences;

Conference Co-Director (Russia) - Dr. Alexander Kaffka, Center for Military Policy, Institute of the USA and Canada, Russian Academy of Sciences; Chairman, CFE Foundation;
Conference Co-Director (USA) - Dr. Kyle Olson, Chemical and Biological Arms Control Institute, Alexandria, VA.

Conference On Sea-Dumped Chemical Munitions
(Kaliningrad, Moscow Region, 12-15 January 1995)

Final Document

The participants of the Conference decide on the following:

1. The old chemical munitions dumped in the North European seas pose a potential environmental danger;

2. The scale of this environmental danger is not qualifiable at the present moment; steps should be undertaken urgently by the international community to provide a solid scientific assessment of the situation, which include the following:

   - conduction of a comprehensive investigation, with the technical means available now, to identify the dumping areas as precisely as possible; conduction of further searches in historical archives to complete the picture of dumping;

   - conduction of repeated on-site investigation of individual dump areas to establish dynamically changing parameters;

   - risk evaluation, including risks of bioaccumulation of chemical warfare agents in biota and potential genetical consequences;

3. Steps should be taken consequently to enable all interested countries to move towards a resolution of this dangerous situation, which should make provision for:

   - selection of appropriate monitoring strategies;

   - marking dumping sites and prioritising them depending on research findings; education of fishermen about proper procedures in case of recovery of individual munitions;

   - design and evaluation of possible remedial strategies.

4. The regular working group on the problem should be established, to coordinate and comment on practical steps and necessary research, until an official international mechanism is elaborated.

Welcoming Address

by M.S. GORBACHEV, President, *Green Cross International*,
former President of the Soviet Union

TO THE PARTICIPANTS OF THE CONFERENCE

on the European Environmental Security - the Problem of Chemical Weapons Accumulated in the War and Post-War Years in the Seas of North Europe.

Ladies and Gentlemen,

I extend greetings to you who gathered here, on Moscow land, to discuss very serious problems.

The very cross-section of the audience here is testifying that the military, environmental and political security issues are very closely interwoven. The ecological aspect of this problem has reached a critical point, growing into a political problem of international importance.

During the decades of the Cold War, all the attempts of its appropriate solution remained practically blocked.

Now such a possibility exists, but it can be realised on the basis of a wide international cooperation only. In this connection, I see as very promising the idea of using the potential of the defence conversion industries for solving the extremely complicated problem of sea-dumped chemical munitions. This idea has been advocated by the Russian conference organiser - "Conversion for the Environment" Foundation.

Your conference is timely and topical. I wish a all the participants successes in your noble course.

M.S. Gorbachev

CONTRIBUTING PAPERS

Section 2.1

General

WAR GASES AND AMMUNITION IN THE POLISH ECONOMIC ZONE OF THE BALTIC SEA

DR. EUGENIUSZ ANDRULEWICZ
Institute of Meteorology and Water Management,
Maritime Branch
Gdynia, Poland

1. Introduction

Information on chemical munition dumped in the Baltic Sea until 1947 was provided by the contracting parties of the Helsinki Convention and observers from the United Kingdom, United States of America and Norway to the Helsinki Commission (HELCOM CHEMU, 1994). According to the submitted information around 40,000 tonnes of chemical munition was dumped in the Baltic Sea after the World War II. It is estimated that the munition may contain about 13,000 tonnes of chemical warfare agents. The following dumping areas in the Baltic Sea were identified (Fig. 1):

- Gotland Deep (south-east of Gotland Deep);
- Bornholm Deep (east of Bornholm);
- Little Belt (southern part of Little Belt).

There are indications that during transport to the dumping area east of Bornholm and south east of Gotland the munitions were partially thrown overboard from ships *en route*. Some munitions were dumped in wooden cases, which might have remained floating for some time and might have drifted outside the intended dumping areas.

In the post war period, until the late fifties, there was a number of reported injuries from suspicious material stranded in the western and middle part of the Polish coast (near Ustka), and also along the Hel Peninsula (near Jurata) (Fig. 2). In newspapers and journals a number of cases about casualties at sea were reported, involving fishermen trawling on bottom with some of them being seriously injured and hospitalised. At that time their injuries were described simply as "burned by yperite".

No recent casualties have been recorded, however, the reports of accidental observations of war material are numerous enough to arise public anxiety. A number of wrecks containing ammunition were located on the sea floor of the Polish Economic Zone (Figs. 2 and 3). The degree of their corrosion is largely unknown. The reports of unidentified bombs, torpedo parts and other military equipment parts caught in fishing nets are still coming up. When lifted up from the sea bottom, these objects are usually thrown back into the sea and not necessarily at the same place. They are often replaced far from where they were caught and sometimes closer to the shore. Fishermen often

A. V. Kaffka (ed.), Sea-Dumped Chemical Weapons: Aspects, Problems and Solutions, 9–15.

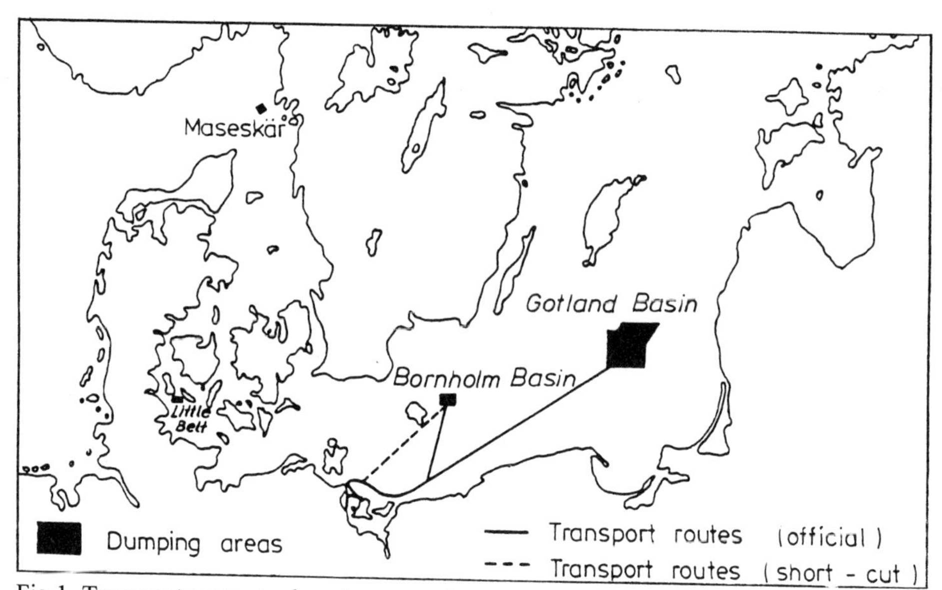

Fig.1. Transport routes to dumping areas (according to HELCOM CHEMU, 1994)

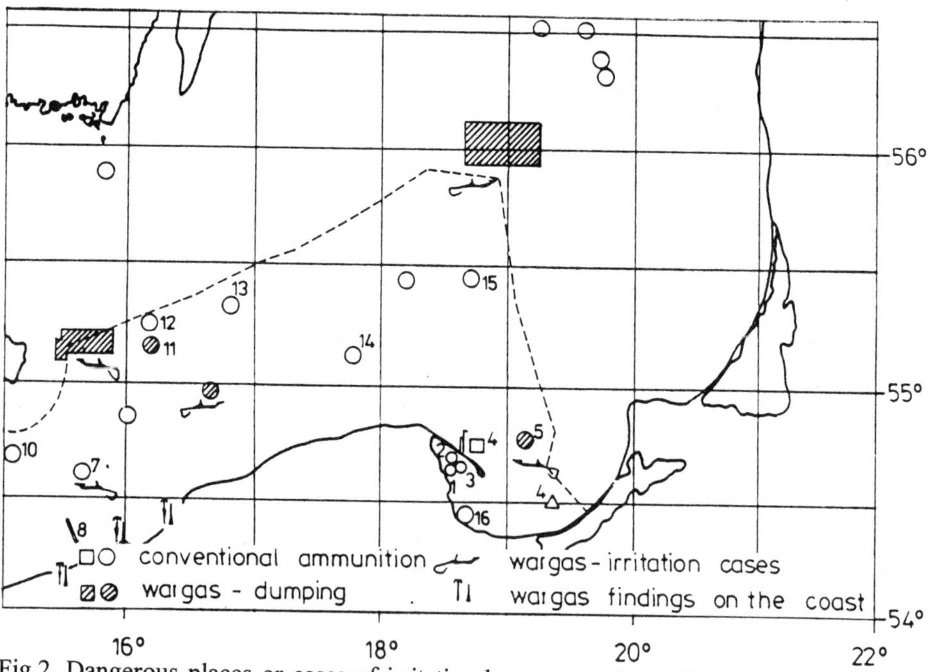

Fig.2. Dangerous places or cases of irritation by war gas (according to Hydrographic Service of the Polish Navy in Gdynia, 1993)

talk about catching unidentified objects that look like war items or barrels containing suspicious material. Most of these cases, for various reasons, are not reported to the appropriate maritime authorities in Poland - the Maritime Offices or the Polish Navy.

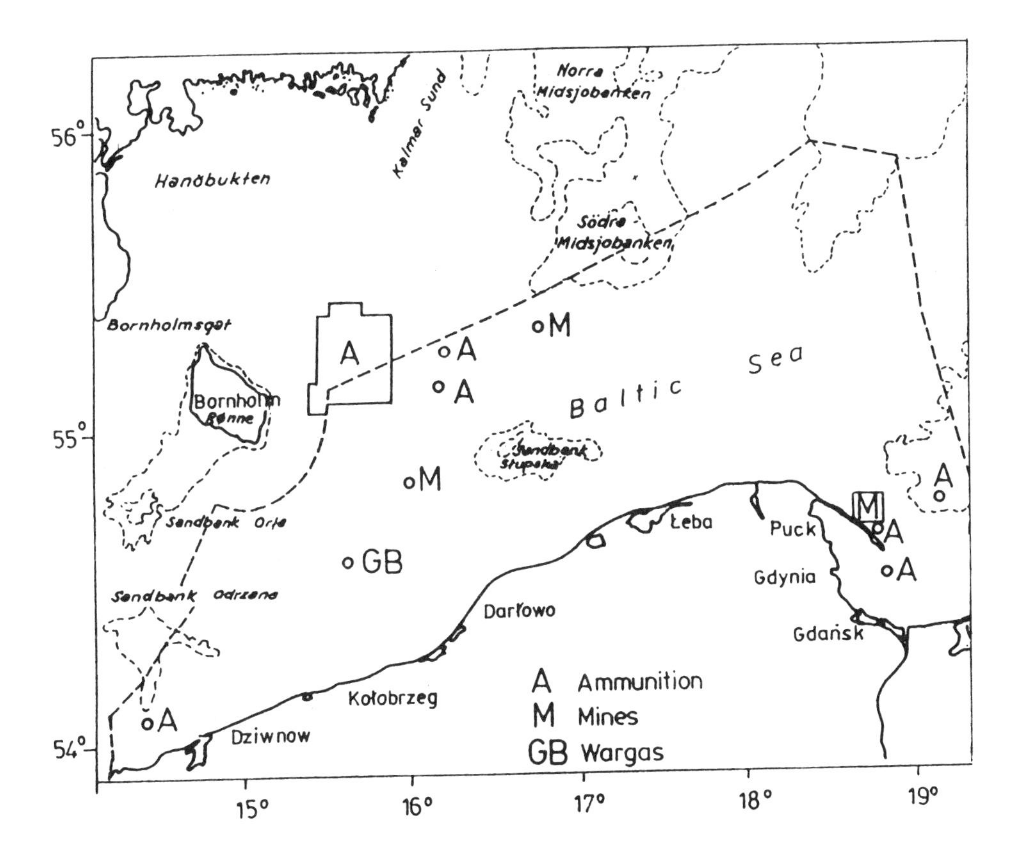

Fig.3. Map of the Polish sector of the Baltic Sea indicating dangerous places (according to Maritime Office in Gdynia, 1993)

2. Information on Dumping Areas and Types of Dumped Chemical and Conventional Warfare Items in the Polish Economic Zone

There is no special service in Poland that records or investigates war items dumped in the sea, however such records have been carried out by the Navy Hydrographic Service in Gdynia, the Maritime Office in Gdynia and the Institute of Meteorology and Water Management -- Maritime Branch in Gdynia.

This information has been submitted to the Helsinki Commission [1]. Information on conventional ammunition, wrecks containing ammunition, dumping areas of chemical warfare agents as well as irritation cases by war gases along the Polish shoreline and within the Polish Economic Zone of the Baltic Sea are included in this report.

The Hydrographic Service of the Polish Navy has identified and marked some of the so-called "dangerous areas" in the Polish Economic Zone [2] (Tab. 1, Fig. 2). The main areas of chemical warfare dumping are observed in the southern part of the Gotland Basin and in the Bornholm Basin (large shaded areas). The quantities of dumped material are unknown.

TABLE 1. Information on conventional ammunition, wrecks containing ammunition and dumping within the Polish Economic Zone of the Baltic Sea according to the Hydrographic Service of the Polish Navy, 1993

No	Coordinates		Comments, year of source
1.	54 40 33	18 34 00	Wreckage of ship with ammunition, 1960
2.	54 42 02	18 37 08	Barge with ammunition, 1959
3.	54 39 06	18 36 48	Trawler with ammunition, 1964
4.	54 29 36	19 25 06	8 wreckages of ships with ammunition, 1957
	54 32 24	19 21 42	
	54 28 30	19 16 54	
5.	54 45 00	19 10 00	Tanker with ammunition, 1969
6.	54 47 00	18 38 00	Numerous spread mines and ammunition
	54 47 00	18 50 00	(Secret German Navigational Information, 1943)
	54 43 30	18 38 00	
	54 43 30	18 50 00	
7.	54 37 06	15 39 00	Ammunition and war gases, 1950
8.	54 23 48	15 31 20	Large-caliber undetonated charges
	54 24 09	15 32 54	Former range of the Kriegsmarine
	54 17 42	15 37 18	German maps from 1948 and 1950
	54 17 18	15 35 42	
10.	54 41 51	15 02 30	Bomb, Danish navigational information, 1991
11.	55 08 00	16 11 00	Ammunition, German navigational maps, 1940s
12.	55 16 00	16 12 00	Ammunition, German navigational maps, 1940s
13.	55 19 30	16 48 00	Ammunition, German navigational maps, 1940s
14.	55 07 42	17 47 30	Tanker, 1955
15.	55 27 21	18 44 30	Barge with a cargo of tar, 1990
16.	54 26 18	18 40 05	150-200 mm charges within the area of an 80-metre radius, 1969

TABLE 2. Dangerous areas due to war items on the bottom of the Polish Economic Zone of the Baltic Sea (The Maritime Office in Gdynia, 1993)

War items	Coordinates
Ammunition	54 45 00 19 10 00
	55 08 00 16 11 30
Ammunition/mines	54 43 30 18 50 00
	54 47 00 18 50 00
	54 47 00 18 38 00
	54 43 30 18 47 08
Mines	55 19 30 16 48 30
	54 51 55 16 48 30
War gas	54 37 00 15 39 00
Wrecks with ammunition	54 32 30 18 50 18
	54 04 54 14 27 18
	54 51 55 16 01 05
	54 42 02 18 47 08

TABLE 3. Chemical munition in the Baltic Sea reported by the *Sailor's Newsletter*, Gdynia, 1948-1970; (Institute of Meteorology and Water Management - Maritime Branch in Gdynia, 1994)

Sailor's Newsletter No.	Coordinates
129/49	55 20 00 15 37 00
141/49	55 35 00 15 42 00
141/49	55 24 00 16 04 00
34/50	54 37 00 15 39 00
54/61	54 48 08 10 13 03
98/69	54 36 00 10 42 00

The Maritime Office in Gdynia has also recorded a number of areas were dangerous "war items" had been located (Tab. 2, Fig. 3). The Institute of Meteorology and Water Management - Maritime Branch in Gdynia (1994) has reviewed the *Sailor's Newsletter* for the period 1946-1947 and has found numerous reports of mines, bombs, wrecks containing conventional ammunition as well as chemical weapon in the southern Baltic Sea. In most of these cases the locations have been identified by the Navy Hydrographic Service and/or Maritime Office in Gdynia, however there is lack of detailed information about other places not mentioned by those institutions. Tab. 3 presents location of chemical munition findings in the southern Baltic Sea.

3. Comments and Conclusions

There is a mounting evidence that the sea floor of the Polish Economic zone is littered with dangerous items due to the post war dumping of war gases and ammunition and a number of wrecks containing ammunition. There is possibility that military exercises have added additional bombs or/and mines to the sea floor, however official records revealing such information do not exist.

Irritation cases and findings of mustard bombs on beaches were recorded mainly in 1952 and 1955. Isolated cases were recorded in 1960s and 1970s. Cases of catching viscous mustard gas or net contamination during bottom trawling were recorded until the end of seventies (1979). Locations of these cases were reported mainly for the western part of the Polish coast which is in good agreement with the available information on dumping locations and dumping routes [3]. Cases of catching barrels containing unidentified chemicals are still recorded. The risk of serious poisoning by chemical agents still exists.

The threat to fish and to fish consumers from war gas is unlikely due to a natural ability of fish to avoid such dangerous or poisonous areas. At present the potential risk towards people has decreased considerably due to the ongoing processes of degradation. There is no evidence of the widespread risk to humans or to the environment from dissolved warfare agents.

The Polish Navy is prepared to handle any reported case of chemical or conventional weapons found on the coast or at sea, however, the risk of bringing single objects/barrels or warfare-looking items from the sea to the shore is bigger than re-dumping. It is therefore recommended to establish a set of guidelines for fishermen and those trawling the sea floor on how to handle dangerous objects and how to report them.

Human risk from conventional weapon is only likely when wrecks with ammunition or the ammunition itself are disturbed on the sea floor. There is no official plans for lifting dangerous items from the sea floor for the purpose of utilising them on the surface.

It is of great interest to the public and to authorities to get more information on this matter. Therefore it might be worthy to undertake a pilot study on contamination of bottom sediments with war gas and about the condition of chemical munition containers as well as on the corrosion degree of conventional ammunition and wrecks containing munition.

References

1. Andrulewicz E. (1993) *National report on war gases and ammunition in the Polish Exclusive Economic Zone of the Baltic Sea*, Ad hoc Working Group on Dumped Chemical Munition (HELCOM CHEMU), Report 2/2/4, Vilnius, Lithuania.
2. *Dangerous places or cases of irritation by war gas in the Polish sector of the Baltic Sea*, (in Polish), The Hydrographic Service of the Polish Navy in Gdynia, 1993, Mimeo.
3. HELCOM CHEMU, 1994, Report on chemical munitions dumped in the Baltic Sea, Rep. 16 Meeting Helsinki Commiss., 8-11 March, 1994.

4. Institute of Meteorology and Water Management - Maritime Branch in Gdynia, 1994, Extract from records on navigation and fishing instructions in the Baltic Sea, Sailor's Newsletter, Gdynia, 1948-1970), Mimeo., (in Polish).
5. The Maritime Office in Gdynia, 1993, Information on dangerous places in the Polish Economic Zone, Mimeo., (in Polish).

PRE-CONVENTION LIQUIDATION OF SOVIET CHEMICAL WEAPONS

DR. LEV A. FEDOROV
Vernadsky Institute of Geo-Chemistry and Analytical Chemistry, Russian Academy of Sciences; Union For Chemical Safety, Moscow

Abstract

A new approach to the problems related to the pre-convention history of Soviet chemical weapons (CW) has been developed. It is focusing on underwater disposal of home-made (Soviet) and captured (German) CW in the seas surrounding Russia and their burial; contamination of numerous territories due to producing, testing and stockpiling CW.

1. Introduction

Official data about the Soviet Union's chemical weapons (CW) is actually confined to information only on those stockpiles stored through the late eighties. It does not give an exhaustive idea of the historical aspect of the Soviet Union's road to the possession of CW. However, it is the analysis of this road that can give a factual basis for solving the problems of overcoming subsequent ecological damage [1].

What we have on the problem of the ecological consequences of previous preparation for chemical war is, in fact, only unofficial information; namely reports by the press and eyewitness recollections.

There are two main groups of questions:

- underwater disposal of home-made (Soviet) and captured (German) CW in the seas surrounding Russia and their burial,
- contamination of numerous territories in connection with the production, testing and stockpiling of CW.

Contamination of the Soviet Union's territory with toxic chemicals (TC) and with the products of their transformation in the environment was begun, however, in reverse order; that is, with the production of CW. Ecological problems were mainly caused by CW that were produced in industrial scale and widely tested, were and are stockpiled in army storage facilities or bases, and were disposed of in large amounts (dumped into seas or buried in the ground).

A. V. Kaffka (ed.), Sea-Dumped Chemical Weapons: Aspects, Problems and Solutions, 17–27.

2. Chemical Arsenal

The Soviet Union's first generation combat CW included:

- stable TC of skin-blistering and general toxic action: yperite (HD) and lewisite (L);
- unstable TC: hydrogen cyanide (AC), phosgene (CG), diphosgene (DP);
- irritating TC: adamsite (DM), diphenylchloroarsine (DA), chloroacetophenone (CN).

Second generation combat CW (highly toxic phosphorous TC) included three types of TC; namely sarin (GB), soman (GD) and V-gas.

Only five first generation combat CW were produced in the SU in industrial scale, namely yperite (HD), lewisite (L), hydrogen cyanide (AC), phosgene (CG) and adamsite (DM). Production was set up in pre-war years and continued throughout the war. Second generation combat CW were produced in large scale beginning in the fifties. In addition, in the sixties, large-scale production of herbicidal weapons was carried out, including components of the notorious combat mixture Agent Orange, viz. 2,4,5-T and 2,4-D [1] .

The SU entered the war with a yperite production capacity of 90,000-100,000 metric tonnes per year and with a lewisite capacity of 12,000 tons. Actually, throughout the entire war, not even one of the yperite and lewisite producing shops was able to reach even its half capacity, so less amounts of TC were produced than had been planned.

There is no complete data on the volumes of yperite and lewisite produced.

It can be supposed that, until the beginning of the war, the army got roughly 6,000-8,000 metric tons of yperite from industry.

During the war years (1940-1945), approximately 120,000 tonnes of first generation TC were produced, including [1]:

- 77,400 tonnes of yperite (HD),
- 20,600 tonnes of lewisite (L),
- 11,100 tonnes of hydrogen cyanide (AC),
- 8,300 tonnes of phosgene (CG),
- 6,100 tonnes of adamsite (DM).

Roughly one third of the yperite and lewisite was charged into munitions, the rest were kept in tanks and barrels. The total amount of chemical munitions produced during the war is 4.5 million units, including:

- 2.6 million chemical mortar shells,
- 1.5 million artillery shells,
- 0.1 million rocket shells,

- 0.3 million aircraft bombs.

From 1958 to 1987 not less than 33,000 metric tonnes of second generation TC were produced, including at least:

- 12,000 tonnes of sarin (GB),
- 5,000 tonnes of soman (GD),
- 15,500 tonnes of V-gas [2].

In 1990-1992, in anticipation of the signing the Convention on Chemical Weapons (CWC), the army presented for inspection and elimination 40,000 tonnes of the current stocks of TC, including:

- 7,700 tonnes of stable TC (yperite, lewisite and yperite-lewisite mixture),
- 32,300 tonnes of phosphorous TC (sarin, soman and V-gas) [3].

There is no information about the fate of the other stocks of CW (i.e. produced, but not declared). Meanwhile, comparison of the data on the production of first generation CW and data on their current stocks indicates a great discrepancy. The same can be said concerning the trophy CW captured in Germany during World War II. That is why the question of the fate of these "missing" CW becomes necessary.

3. Ways of Pre-Convention Disposal of CW

Let us analyse ways of taking CW out of circulation.

CW which became needless and/or lost their combat capabilities were disposed of in two ways. Disposal was carried out most intensely in three different periods [1].

The ways of TC disposal included:

- dumping munitions charged with TC into seas (roughly 1/3 to 1/2 of the amount produced during the war);
- land burial of the remaining 2/3 of the TC, which were not charged into munitions.

<u>The first</u>, most intense wave of CW disposal occurred immediately after World War II from 1946 to 1948. In this period, along with the underwater disposal of trophy CW, the army carried out liquidation of stocks of home-made weapons. The amounts of home-made TC liquidated in this period are as follows: roughly 60,000- 65,000 tonnes of yperite and approximately 14,000 tonnes of lewisite. The amounts of the trophy TC liquidated are reported in [16].

<u>The second</u> wave of CW disposal occurred between 1956 and 1962, the period of conversion from first-generation to second-generation CW. Approximately 10,000-

12,000 tonnes of home-made yperite were liquidated. There is no data about the amounts of trophy CW subjected to underwater disposal during these years.

The third wave was in the eighties. The Soviet Army carried out large-scale transfers of CW from 1987 to 1989 from the bases where they had been stored to seven bases which were subsequently announced. At the same time the arsenals were "erased" to dimensions that would be comparable with US stockpiles.

Depending on the character and departmental jurisdiction, several methods of CW disposal were used [1]:

- incineration and/or burial of large amounts of TC not charged into munitions in the vicinity of storage bases and testing grounds in the former SU;
- incineration and burial of TC, including those charged into munitions, in the vicinity of CW production plants in Chapaevsk, Dzerzhinsk, etc.
- underwater disposal (dumping) of chemical munitions and containers filled with home-made and trophy TC in seas; there are indications of dumping chemical munitions into rivers and swamps also.

4. After-Effects of Production

Large-scale production of TC in the SU was primarily carried out in the Volga basin. The waters of the Volga, Oka and Kama were intended to play a twofold role; namely, to serve production purposes and absorb effluents [1].

Before World War II the capacity for large-scale production of first-generation TC existed in 9 cities, while actual production of CW took place in only 5 cities. Russia's current stocks of phosphorous TC are associated with the work of two plants, namely, an old plant in Stalingrad (Volgograd) and a newly built plant in Novocheboksarsk (Chuvash Republic).

A few examples demonstrating the scale of ecological consequences of CW production are given below.

CHAPAEVSK (Samara Oblast). In this town situated on the shore of the Chapaevka River, not far from its confluence with the Volga, the first large-scale yperite production was organised. In pre-war years, the plant occasionally produced batches of yperite and lewisite. Continuous production of yperite and lewisite at the plant in Chapaevsk was carried out only during war time. Yperite was produced up until 1943, after which its production was stopped as it was impossible to prevent injuries to people. Lewisite production carried on throughout the war. Munitions were charged with yperite, lewisite and mixtures of the two at this plant, at the expense of imported raw materials [1].

The air tightness of the technological equipment in the plant was not ensured. Air from shops that produced yperite and lewisite was expelled directly into the town's atmosphere without purification. Waste water purification facilities as a rule did not function. Waste was dumped directly into the Chapaevka River and from there reached the Volga. Spoilage (TC and munitions) went to a dump, now forgotten [1], on the territory of the plant. After the war's closure, considerable amounts of CW rejected by the army up to that time were destroyed directly on the territory of the plant. The plant's own storage facility was eliminated in the late fifties.

There is no data about the degree to which the plant and the town were contaminated by yperite and lewisite (and, generally, by arsenic). However, ecological examination carried out in the winter of 1993-1994 showed that the concentration of arsenic in the soil on the territory of the plant exceeds maximum permissible concentration (M.P.C.) by a factor of 8,500 [4]. In the town's areas surrounding the plant the concentration of arsenic exceeds M.P.C. by a factor of 2-10 times. As for lewisite, it too has not yet vanished. In any case, today, 50 years after the halt of production, a product of lewsite's hydrolysis, even more toxic than lewisite itself, was found in the soil on the territory of the plant.

DZERZHINSK (Nizhniy Novgorod Oblast). Here, on the shore of the Oka at what is now "Kaprolaktam", production of Levinstein yperite was set up in 1939, and continued until 1942 when the plant was converted to the production of Zaykov yperite, which was then produced until the end of the war. This same plant was involved in lewisite production in the war and post-war years. Purification of yperite-contaminated air was ineffective and such air spread over the densely-populated residential areas within a 5-7 km radius of the plant. In the years of intensive production of TC, effluents from the chemical plants passed to the Oka through a system of lakes [1]. Barrels used for packaging arsenic were buried directly on the territory of the plant.

STALINGRAD (VOLGOGRAD). Yperite production was organised on the shore of the Volga in the early thirties. In pre-war years individual lots of yperite were produced. During the war, production of yperite and aircraft bombs charged with the chemical was carried out until the autumn of 1942, and was not restarted afterwards. Experimental production of sarin, soman and V-gas was set up in the forties and fifties. Large-scale production of sarin was started in 1959, and soman in 1967 [1].

In 1965, as a result of damage to the dam, contents of a storage unit where effluents of shops producing phosphorous TC had accumulated for many years broke through into the Volga. The surface of the river was white with a flood of dead fish as far as Astrakhan [5]. The ecological consequences of this and other accidents were not evaluated.

BERESNIKI (Perm Oblast). Yperite production organised at a soda plant in Berezniki (on the Shore of the Kama) was carried out for three years of the war. Units for purification of the effluents from the yperite plant were not built [6].

KINESHMA (Ivanovo Oblast). Production of adamsite and diphenilchlorarsine was set up in pre-war years in Kineshma, at Aniline Ink Plant, on the shore of the Volga. In reality, only adamsite was produced there [1].

NOVOCHEBOKSARSK (Chuvash Republic). Industrial production of V-gas was set up in 1972 at the "Khimprom" plant on the shore of the Volga. The plant was built specifically for this purpose. Charging munitions was carried out until 1987. The wastes of V-gas production went to an unequipped disposal site, and effluents were dumped into the Tsivil River whence they entered the Volga. There is no data about the methods for their purification. Spoiled munitions were destroyed directly on the territory of the plant. **Among significant accidents at the plant, a special place goes to the fire at the storage facility for aircraft bombs charged with V-gas, which occurred on 28 April 1974.** Accidents with release of V-gas occurred even after this [1,5] .

A full-scale analysis of the environmental impact of 15-years of CW production was not made. However, the consequences can be estimated by the following fact: over this period there was a 2-km sanitary safety zone around the plant in which people were not allowed to live or cultivate agricultural products. In the summer of 1994, seven years after the end of TC production, this safety zone was expanded to 4 km [7].

5. Burial

For international inspection, which is expected after the ratification of the CWC, Russia is presenting 2 types of first-generation TC, namely yperite, lewisite and their mixture (in amounts of 7,000-8,000 tonnes, mainly in tanks), and 3 types of phosphorous TC (in the amount of 32,000 tonnes, all in munitions).

Data about the fate of all yperite and lewisite produced in the SU (outside of a known, insignificant amount) is not available. Similarly, there is no information about the fate of the stocks of all hydrogen cyanide and phosgene produced. Irritants (for example, adamsite) excluded from the CWC at the last moment, as well as herbicides not considered within the framework of the CWC at all, are not subject to inspection. Burials of CW carried out from 1946 through January 1, 1977 (land burials) and January 1, 1985 (sea dumping) are also not subject to inspection.

With regard to the large-scale burial of CW, there is data on storage facilities and testing grounds that were situated near Chapaevsk, Kambarka, Gorniy, Shikhany, Arys' (the south Kazakhstan) and so forth.

<u>A few examples.</u>

Beginning with the war, CW produced by the plants of Chapaevsk, Dzerzhinsk and other towns, were kept on a storage ground near Chapaevsk (Pokrovka settlement). In the late forties, tanks with unnecessary yperite were transported to Arys' [8,9]. In the early sixties, at least 1,200 tonnes of yperite were disposed of by means of incineration and burial, while bombs charged with lewisite were sent off for underwater disposal in northern seas [10].

At the Central Military Chemical Testing Ground (TsVKhP), which in 1926 was lodged on the shore of the Volga in Shikhany (Saratov Oblast), 3,200 tonnes of adamsite were buried in the sixties. At present, 3,400 tonnes of irritants are kept at a CW storage facility in TsVKhP. **In 1992-1994 CW were destroyed on TsVKhP by exploding munitions in the open** [9].

CW were tested and stored at many sites around the country. For instance, in Moscow, testing, stockpiling and burying of CW were carried out at what is now Kuzminki [1,9]. In connection with similar activities, high contamination of Moscow's territory, at what is now Ochakovo, was noted as far back as pre-war years.

The same applies to the Nizhniy Novgorod (Florishchi and Gorokhovets camps) and St. Petersburg (Luga, etc.) regions, as well as many others.

Data about the Soviet Union's pre-war CW storage sites and bases was little known by intelligence. Accordingly, without specific historical information, it is difficult to evaluate the possible ecological damage. For example, it turned out that a well-known list of pre-war CW storage sites [11] did not include the Svobodnyy settlement (Amur Oblast) where yperite had been stored for many years. In spite of the fact that strategic missiles are on guard there at present and Russia's spacecraft launching ground will be housed there in the future, there is no data showing that the territory was cleaned up in the aftermath of CW stockpiling. The same applies to the Tatishchevo settlement (Saratov Oblast) where CW were stored in pre-war years and strategic missiles are located at present. The fate of a large military range in Totskiy (Orenburg Oblast) should be considered in a similar way: CW were tested on the range before 1954 when the first exercises involving the use of nuclear weaponry were carried out there.

New storage sites were created during the war. At present, only some of these former (now abandoned) storage sites and bases are known. Ecological evaluation of CW testing grounds and storage facilities was not made.

At present the existence of 7 specialised arsenals of CW are declared (there is an eighth stockpile with irritants at TsVKhP in Shikhany) [1,9].

Two storage facilities belong to chemical forces; those in Kambarka (Udmurtia) and Gorniy (Saratov Oblast).

The storage base in Kambarka was founded in 1941. At present, lewisite is stored here in tanks (6,400 metric tonnes, 80 tonnes in each of 80 tanks). In the mid-50s and in the sixties, yperite and lewisite were destroyed here [1]. Large amounts of yperite (thousands of tonnes) were destroyed by means of incineration in the open air. There was an attempt to destroy lewisite by incineration in the open air, too.

The storage base in Gorniy was founded in 1943. Skin-blistering TC (yperite, lewisite and a mixture of the two) are stored here. Yperite and lewisite were destroyed at the base in the late fifties [1]. Part of the chemical munitions were sent for underwater disposal in the Sea of Okhotsk [12].

Two bases belong to GRAU (Main Missile and Artillery Administration): in Shchuchye (Kurgan Oblast) and Kizner (Udmurtia). The storage base at Kizner has existed since 1914. Its chemical munitions arsenal of rocket and non-rocket artillery evolved between 1954 and 1987. Munitions stored here are charged with phosphorous TC and lewisite. Related to the Kizner settlement is an accident that occurred at the beginning of 1993 on a railroad not far from the large storage facility where CW and other munitions were stored [13]. The arsenal at Shchuchye stores missile warheads and rocket and non-rocket ammunition charged with phosphorous TC. Conventional artillery arms are also stored on the same premises. Related to the base in Shchuchye, a sharp increase in the mortality rate among children under the age of 2 was detected in the district of Shchuchye in 1989 (in comparison with other rural districts of Kurgan Oblast). Subsequently this index returned to normal [14].

The other three CW storage bases belong to VVS (Air Force): in Leonidovka (Penza Oblast), Maradykovskiy (Kirov Oblast) and in Pochep (Bryansk Oblast). Stored at these storage bases are aircraft bombs, universal container units and spray rigs charged with phosphorous TC. The base in Leonidovka has functioned since 1939. A big fire that occurred at the base in 1977 is well-known. Aircraft chemical munitions began being delivered there in the seventies (also stockpiled there are munitions with yperite-lewisite mixture). Munitions charged with phosphorous TC were once disposed of at the base by dumping them into a swamp. Before dumping, the munitions were shot with machine-guns [15]. The base in Pochep is located at one of few sites of the region which were untouched by the "Chernobyl spot".

6. Dumping

Official information about dumping CW into seas and oceans is minimal. For the time being, there is either denial, or forced acknowledgment, which, however, is not accompanied by facts. For instance, there is a well-known letter, written by Danilov-Daniljan, the Minister for the Protection of the Environment and Natural Resources, to the Security Council of RF: "The Ministry, ... having collected, analysed and summarised the information connected with the problem of underwater disposal of CW in Russia's sea areas, arrived at the preliminary conclusion that such anthropogenic pressure had been subjected to the Baltic, White, Barents, Black Seas and the seas of Okhotsk and Japan over a period of more than 50 years" [9].

One can try, however, to get a more detailed picture.

The points of departure whence chemical munitions were transported for underwater disposal were numerous land storage sites: Chapaevsk [8], Leonidovka [1], Gorniy [12], Obozerskiy (Arkhangelsk Oblast) and so on. The nomenclature mainly consisted of Soviet munitions; however, a considerable share of it comprised Soviet-German munitions (i.e. those captured in Germany and included in the stocks of the Red Army). Transhipment to ships was carried out at the respective ports:

- at Liinakhamari port (Murmansk Oblast) by motor transport from Pechenega station for subsequent dumping in the Barents and Kara Seas,
- at a military port in Severodvinsk for subsequent dumping in the White Sea,

- at Posyet and Nakhodka ports (Primorskiy Krai) for subsequent dumping in the Sea of Japan,
- at Paldiski and Tallinn ports for subsequent dumping in the Baltic Sea and so forth.

Transportation was carried out by naval ships in close co-operation with the merchant marine. It is known that many accidents occurred in which participants on these expeditions were poisoned by TC as a result of leaks in the tightness of the chemical munitions during loading and unloading.

As for specific dumping sites, information about them is not as inaccessible. It can be found, for instance, on navigation maps where these sites were ciphered as "dumps of explosive substances". For example, two such sites, mentioned by eye-witnesses, were indicated on a map of the White Sea; a site in the Kara Sea, near the northern part of Novaya Zemlya, and a site near the Povorotnyy Cape in the Sea of Japan. In total, there are no less than 12 large sea areas, although there are hundreds of specific sites [9].

A few words about facts and documents. One of rare exceptions to the information blockade that has existed for many years is a report on underwater disposal of trophy (German) CW which was carried out by the Soviet Navy from 1946 to 1948 in the Baltic Sea [16]. It is worth supplementing this report with a vivid document concerning the same issue taken from "Stalin's Special Folder," which was recently declassified [17]:

Top secret

2 July 1948
N 3215/K

To Comrade Stalin I.V.

At the beginning of June of this year three crews of fishing boats belonging to the base of active fishery of the Ministry of Fish Industry of Lithuanian SSR carried out fishing 70-75 miles west of Klaipeda town. Three 250-kg German aircraft chemical bombs charged with yperite were lifted on board with the seine from a depth of 100-120 meters <...>

In June of this year, 9 fishermen from fishing boat ZMR-93 "Bauska" were brought to the city hospital in Liepaja, Latvian SSR with signs of poisoning by yperite.

A preliminary investigation showed that the fishermen, being in the sea on 9 June at a distance of 40-50 kilometres from Palanga, fished out a 250-kg box with an inscription in German. When the box was opened an aircraft bomb was found. It had a fractured head and was producing a strong smell.

The box with the aircraft bomb was dropped into the sea <...>.

In accordance with the data available at the Ministry of Internal Affairs of the USSR, in 1945, our troops captured 35 [thousand - L.F.] metric tonnes of chemical weapons, including chemical aircraft bombs, which were brought and dumped by the trophy groups into the sea between the coasts of the Soviet Union and Sweden.

Measures are being taken to detect other bombs.

Minister of Internal Affairs *S.Kruglov.*

7. Conclusions

In chemical disarmament, the interests and attitudes of the army and the population do not coincide. The army does not differentiate between the combat and ecotoxic characteristics of TC. Thereby the army simplifies the process of avoiding chemical confrontation by reducing it to the mere disposal of existing stocks of CW. However, the ecotoxic characteristics of TC are of importance to the health of the population and ecology, even when they are no longer of any value for combat.

The methods used in the pre-convention elimination of CW were ecologically hazardous. The consequences of such "disposal" will be making themselves evident for a long time. Thus all CW that were produced and liquidated have turned from combat weapons into ecological ones.

Elimination of existing stocks of CW within the framework of the CWC resolves only the international (military and political) aspect of the problem. The aspects connected with the medical and ecological after-effects of preparation for chemical war bear a domestic character and are, as a rule, not discussed. There is no state body in Russia is responsible for solving the overall problem of overcoming the ecological and medical consequences of preparation for chemical war.

The above account shows contemporary Russia's evident and fundamental problem with chemical disarmament. It is necessary to formulate and solve not one, but two nation-wide problems in ecological safety in ending the CW confrontation:

- analysis of the military-chemical past, and development of work on overcoming ecological and other consequences of preparation for chemical war,
- ecologically safe elimination of CW.

Being of nation-wide importance, these two problems must be addressed simultaneously, although in principle they may be solved at different paces and possibly by different programs.

References

1. Fedorov, L.A. (1994) *Chemical Weapons in Russia: History, Ecology, Politics,* Center of Ecological Policy of Russia, Moscow.
2. *Rossiya,* 8 December 1993.
3. Draft Comprehensive Program of Staged Destruction of Chemical Weapons in the Russian Federation. Phase 1. Priority Measures on Preparing to Meet International Obligations in the Area of Destroying Chemical Weapons Stockpiles, Moscow, 1992.
4. *Segodnya* (Moscow), 11 February 1994.
5. *Izvestiya*, 2 December 1992.
6. *Segodnya* (Moscow), 10 August 1993.
7. *Sovetskaya Chuvashiya*, 3 August 1994.
8. *Chapayevskiy Rabochiy*, 14 May 1993.
9. *Obshchaya Gazeta,* 16 September 1994.
10. *Chapayevskiy Rabochiy*,, 17 October 1992.
11. Krause, J., Mallory, C.K. (1992) *Chemical weapons in Soviet military doctrine. Military and historical experience, 1915-1991*, Westview Press, Boulder.
12. *Saratov*, 20 March 1993.
13. *Vek* (Moscow), N 27, 1993.
14. *Segodnya* (Moscow), 5 July 1994.
15. *Obshchaya Gazeta,* (Moscow), 11 March 1994.
16. *Comprehensive Analysis of the Danger of Captured German Chemical Weapons Disposed of in the Baltic Sea During the Post-War Period*, Research Report, Moscow, Military Unit No 64518, 1992.
17. GARF. f. 9401, OP. 2, D. 200b L.306-307.

CHEMICAL "ECHO" OF THE WARS

A.V. FOKIN, K.K. BABIEVSKY
Nesmeyanov Institute of Elementorganic Compounds
Russian Academy of Sciences, Moscow, Russia

The general public underestimates the hazard of chemical weapons, to a great extent, due to insufficient knowledge of the action of this means of mass destruction. According to the terms and conditions of the Chemical Weapons Convention, states possessing stockpiles of chemical weapons must have them destroyed early in the 21 century. According to the experts' estimations, these stockpiles (in terms of the mass of chemical warfare agent) include more than 30,000 tonnes in the USA and 40,000 tonnes in Russia.

Such large-scale destruction operations proved to raise complicated economic, social and ecological problems, and may yet have profound effects on population and the environment. The dumpings of old chemical warfare munitions and repositories of CW agents, remaining after past wars, still further complicate things.

Chemical weapons became the first means of mass destruction to be used in combat operations. They demonstrated high effectiveness in comparison with conventional weapons and had an essential influence on the course of fighting in World War I. On April 22, 1915, German forces marked the beginning of the chemical war with a gas attack of chlorine sweeping out from cylinders near Ypres in Belgium. As a result of the attack, 15,000 people were injured in French Army positions: 5,000 of them died within the following 1-2 days. On May 31, 1915, German forces again conducted a gas attack, this time on the Eastern front in the region of Bolimov. The attack resulted in losses in the Russian Army exceeding 9,000 causalities of chemical warfare agents, including 1,200 fatalities. The total amount of CW agents, produced by chemical industries of the belligerent parties during World War I, was more than 130,000 tonnes. The main part of these CW was used on battlefields or concealed in European land in non-exploded shells.

In 1936 during the aggressive war in Abyssinia, fascist Italy used chemical weapons (especially mustard gas and phosgene) not only against the military forces, but also against civilian population. As a result, the number of chemical war casualties increased by 15,000.

From 1937 to 1943 the Japanese Armed Forces used chemical weapons in China; the Chinese civilian population also suffered.

During World War II, military R & D centres in the leading industrial countries continued to work intensively creating both new CW agents and technical facilities for

A. V. Kaffka (ed.), Sea-Dumped Chemical Weapons: Aspects, Problems and Solutions, 29–33.

their delivery. Along with conducting research, fascist Germany promptly developed a powerful industrial basis for the production of CW agents and various kinds of chemical munitions. As captured German archives show, in 1943 the annual capacity of plants producing CW agents was almost 180,000 tonnes which was 1,5 times greater than the amount manufactured by all belligerent countries during World War I. According to the evidence of A.Speer, Hitlerite Minister of Industry, at the Nuremberg Trial of German war criminals, three specially designed plants were built to produce new CW neuroparalyzant agents (tabun and sarin), and they had been working effectively for the entire period of the war. One of these plants, built in 1942 on the territory of occupied Poland in Dihenfurth (near Wroclaw), alone produced 12,000 tonnes of tabun in the war years. The production of the more toxic CW agent - sarin - was performed at an industrial installation in Falkenhagen, near Fürstenberg (500 tonnes per year). In addition, German chemical facilities also manufactured other CW agents in large scale, specifically, "Cyclone B" (hydrocyanic acid adsorbed on powders), used by Nazi for the mass murder of people in the gas chambers of extermination camps. Toward the end of the war, in Auschwitz alone, 4,5 million prisoners were killed by this agent. Fascists also used chemical weapons against Soviet partisans on temporarily occupied territories, in particular, in May-June, 1942, in the Adzhimushkaiskie stone quarries in the Crimea and outside Odessa.

Throughout almost the entire war, Nazi's kept the world in fear of massive military use of chemical weapons. The High Supreme Command of Wehrmacht conducted direct preparations for the use of CW against the USSR just prior to their attack on the USSR. On June 11, 1942, German forces received confidential instructions concerning the waging of chemical war, which emphasised the necessity of suddenness and the large scale of CW use by all kinds of troops. Then, the Nazi's began to move in and stockpile chemical warfare munitions on the occupied Soviet territory. As is known from an article published in the daily *Krasnaya Zvezda*, just a few years ago one of these German stockpiles of artillery shells filled with CW agents was found in Belarus and detoxified by the military chemists.

In spite of all this, Hitlerite Armed Forces did not unleash large-scale chemical warfare. Undoubtedly, the high preparedness and gas alert readiness in the Soviet Union, the fear of a retaliatory strike with the same weapons by the anti-Hitler coalition affected this decision. The fascists knew that the Red Army had had special units of chemical protection and efficient technical and medical facilities for defence. The work of our outstanding chemist Academician N.D. Zelinsky and his school had a determining influence on the development of anti-gas protection, and all veterans of the Great Patriotic War remember well our reliable army gas masks.

After the crushing defeat of Germany, the allies found more than 250,000 tonnes of aerial bombs, artillery shells, mines and grenades filled with mustard gas, phosgene and organoarsenic CW agents, as well as neuroparalyzant agent (tabun and sarin), in German chemical arsenals. Considerable amounts of CW agents were also found in industrial reservoirs and special containers.

The stocks of "quite death" were great enough to kill or make disabled millions of people. The allies faced an urgent problem: what should they do with these "arsenals of death". After hurried consultations the allies made the decision to establish a special organisation (which is known as "Continental Committee on Dumping") in order to co-ordinate operations on the disposal of chemical weapons. They also decided that the occupational authorities of every zone would destroy the CW stocks found on their territory by their own forces according to their own plans, thus they could act in the manner most convenient for them. The operations of destruction were performed hastily, under conditions of strict confidence, and, as became evident, with neglect for ecological safety standards.

However, the undeniable fact should be noted: the majority of the chemical weapons produced in fascist Germany fell into hands of British and US Armed forces. Until now, the States-participants in the "Continental Committee" have not published any official information on what they did with captured German, or their own, CW manufactured during World War II (in the course of the Helsinki Commission's activities Germany, Russia, the UK, and the USA provided official reports on dumping in the Baltic and the North seas - *Editor.*) According to the results of a special investigation held by the Stockholm International Peace Research Institute (SIPRI), published in 1971 and 1978, several thousand tonnes of mustard gas, which US Forces found in their zone, were destroyed by on-site burning (it is terrible to image how much of the notorious dioxin was formed in this case); part of CW was transported out of German territory, and almost all remaining CW were dumped in the sea. The allies chose the Baltic Sea and the North Sea (which are fairy shallow) as the main sites for dumping chemical warfare munitions in Europe; dozens of military and civilian ships were used to conduct these operations.

Within the period of 1945-1949, Great Britain sunk barges with 175,000 tonnes of German CW munitions as well as those of its own at a site located 20 miles to the west of the coast of Ireland. From 1955 to 1957 these operations were continued. Thus, for example, 25,000 tonnes of captured German aerial bombs, filled with tabun, and British bombs and shells with mustard gas and phosgene were dumped at a site located at 56^0 30' north latitude and 12^0 west longitude, 250 miles to the west of Coloncey Island (the Internal Hebrides). In addition, as contemporary newspapers reported, in 1965 approximately 1,700 barrels of mustard gas were sunk in the Bay of Biscay at a depth of about 2 km.

The Baltic Sea has not been lucky either. As soon as the war was over, more than 200,000 tonnes of captured chemical weapons were accumulated in the Baltic ports of Kiel, Emden and Wolgast, and sent for dumping on the initiative of the "Continental Committee". In his article, Swedish journalist Frederick Laurin [3] cited the facts, told by a participant of the above operations, who in 1946 was a deck-hand on the "Monte Pascoal", a damaged German passenger ship. "The ship was loaded from barges with some 3,000 tonnes of CW ammo. ... When we got to Kiel, charges were mounted on the inside of the hull by American engineers. Concrete was put into the holds and the cargo doors welded together. Then we were towed by tugs from Kiel to Skagen, and on the twenty-first of December ... the charges went off and she went down slowly. It was about 20 nautical miles from Skagen". Only a few of the German workers, who loaded the ship, remain alive. Several workers were poisoned heavily and died, many of them lost their health.

The hydrographic services of European countries mark some dumping sites as regions of potential navigational or ecological hazard. But these designations do not give an idea of the scale and true character of this threat. A Swedish investigator of the Baltic Sea, Bjorn Åkerlund, believes that the largest dumpings of German chemical weapons from World War II are in Skagerrak region. The remains of ships destroyed by corrosion, which contained up to 180,000 tonnes of mustard gas in their concrete holds, are located at this site, 20 miles to the west of the Swedish port Lysekil at a depth of about 200 m. There are German motor torpedo boats (T-38 and T-37), mine sweepers (M-16 and M-522), as well as the former passenger liner "Horn" among these ships. US servicemen oversaw the execution of the ships' dumping. In 1990 in the same region, opposite the Norwegian port of Arendal, a group of Norwegian submarines found 15 captured German ships with chemical weapons which had been dumped on May 18, 1946. The British weekly *Sunday Times Magazine* of May 5, 1992 [2], published the memoirs of Dr. Stephen Musgrave, who served at the British Naval base in occupied Kiel at that time: "Our information was that some 500,000 tonnes of poison gas of various types were to be taken out of Kiel to Emden and disposed of. I personally watched vessels being loaded with a wide variety of shells and canisters. The vessels were then towed north to the Skagerrak and scuttled in such a way that they sank horizontally, reducing the risk of a highly dangerous cargo shifting and igniting". One additional site of dumping is known, namely in the Lilla Bält, or Little Belt (between the Danish island Fyn and the European continent), where ships with about 5,000 tonnes of the nerve agent tabun in their holds were sent to the sea bottom.

In 1985, taking into account the great public interest in the CW dumpings found in Skagerrak, Norway prepared a comprehensive report on the results of investigations conducted in the region by the National Defence Research Establishment. On June 26, 1990 the report's summary was submitted to the Conference on Disarmament in Geneva as a working document. But two days later it was withdrawn. The journalists who dealt with the Conference were informed that this report was removed after the Western countries had put pressure on Norway's Foreign Ministry.

At present, as the sites of some CW dumpings of the passed war are becoming known, the question arises: should these chemical weapons be lifted from the bottom and detoxicated in a proper way or not? Many experts believe that such a decision is late and, therefore, hazardous. For many years the dumped ships have been decaying under the action of sea water, the shells of chemical munitions are also decaying. Undoubtedly, the consequences of the careless treatment of these stockpiles are very dangerous for people and the environment. In many regions of the Baltic Sea, cases of heavy chemical poisoning were registered among fisherman having found shells and containers with CW agents in their nets. However, ecologists reason that along with the hazard of direct contact injuries, danger may consist in the slow, but prolonged influence of super-toxic compounds released from destroyed shells of the sunken CW munitions on the biosphere.

The remains of CW munitions and canisters containing mustard gas are the most hazardous. This CW agent has manifold injurious capabilities connected with its ability to inhibit the enzyme systems of living organisms which results in the termination of the intracell metabolism and the necrosis of tissues. This CW agent is not actually soluble in water, it only becomes crusted with a jelleous coating consisting of the products of its incomplete destruction and polymerisation, which prevents the further hydrolysis of the CW agent. Due to this, mustard gas placed under the water can conserve its toxicity for decades. The organophosphorus poisons - tabun and sarin - are slowly hydrolysed in water. A few reports appeared in the US press saying that the intermediate products of these substances' hydrolysis have almost the same toxicity as the CW agents themselves, but they are 100 times more stable to further destruction.

It is easy to suggest what happened with the sunk shells containing phosgene. This compound reacts with water rather quickly, giving harmless products. Things are more complicated with organoarsenic CW agents. For instance, adamsite does not change and maintains its toxic properties even after staying under water for many years. On the contrary, lewisite is rapidly hydrolysed in water, but it gives other organoarsenic products which are toxic.

Unfortunately, nobody now knows clearly what is occurring with the dumped chemical weapons at the sea bottom and what should be done with them. Undoubtedly, further inactivity can not be tolerated. It is necessary to expediently unite the efforts of military forces and scientists in order to prepare a total list of the CW dumping sites, and to obtain a clear picture of the situation in the dump zones of the Baltic Sea.

Only thereafter shall we be able to assess the true danger and to work out specific measures to prevent (or to reduce) hazardous ecological consequences.

References and Literature

1. Fokin, A.V. and Babievsky, K.K. (1992) The destruction of chemical warfare agents, *Priroda (Nature)* **5**, 16-25.
2. Knightley, P. Dumps of Death, *The Sunday Times Magazine*, May 5, 1992, 27-30.
3. Laurin, F. (1991) Scandinavian underwater time bomb, *The Bulletin of the Atomic Scientists* **2**, (**47**), 10-15.
4. Lundin, J. (1979) Stockpiles of chemical weapons and their destruction. *SIPRI Yearbook of World Armaments and Disarmament*, 470-489.
5. Robinson, J.P. (1983) in: *SIPRI Yearbook of World Armaments and Disarmament*, 409.

SPECIAL STUDY ON THE SEA DISPOSAL OF CHEMICAL MUNITIONS BY THE UNITED STATES

MARK J. FRONDORF
Science Applications International Corporation
McLean, Virginia, USA

1. Introduction

SAIC undertook a special study on the sea disposal of chemical weapons at the bequest of the U.S. Arms Control and Disarmament Agency. This study was commissioned for the purpose of obtaining the most comprehensive and authoritative historical account possible regarding U.S. sea disposal of chemical weapons. What follows is a summary of the report with special emphasis on the dumping that took place in the North Sea and Skagerrak Strait.

After thoroughly searching records and files involving the dumping of chemical weapons at sea, 60 U.S. sea dump sites located around the world were identified, several of which were used more than once. Although a number of them were created specifically to dispose of W.W.II CW stock, no evidence was found of any U.S. dumping in the Baltic Sea, during, immediately after, or any time since W.W.II.

2. “Operation Davey Jones Locker"

The United States dumped chemical munitions in Scandinavia during "Operation Davey Jones Locker" from June 1946 through August 1948. Eleven ships were scuttled in this operation: nine ships in the Skagerrak Strait and two hulks in the North Sea. The Skagerrak Strait is located between the North Sea and the Baltic Sea. During this timeframe, the Soviet Union was using the Baltic Sea to dispose of W.W.II CW stock.

At the conclusion of World War II, all captured chemical weapons within the United States zone of Germany were transferred to five former German ammunition depots. They were: Frankenberg; Wildflecken; Grafenwöhr; Schierling; and, St. Georgen. From these five depots U.S. demilitarisation operations were undertaken employing chemical decontamination, detonation, burning, and scuttling at sea. In addition, a fair amount of CW stock was shipped to the United States.

Figure 1 is a map taken from *The History of Captured Enemy Toxic Munitions in the American Zone European Theater May 1945 to June 1947* identifying the location of the five German depots. Figure 2 is also from the same document and identifies the locations of the first five scuttlings under Operation Davey Jones Locker.

A. V. Kaffka (ed.), Sea-Dumped Chemical Weapons: Aspects, Problems and Solutions, 35–40.

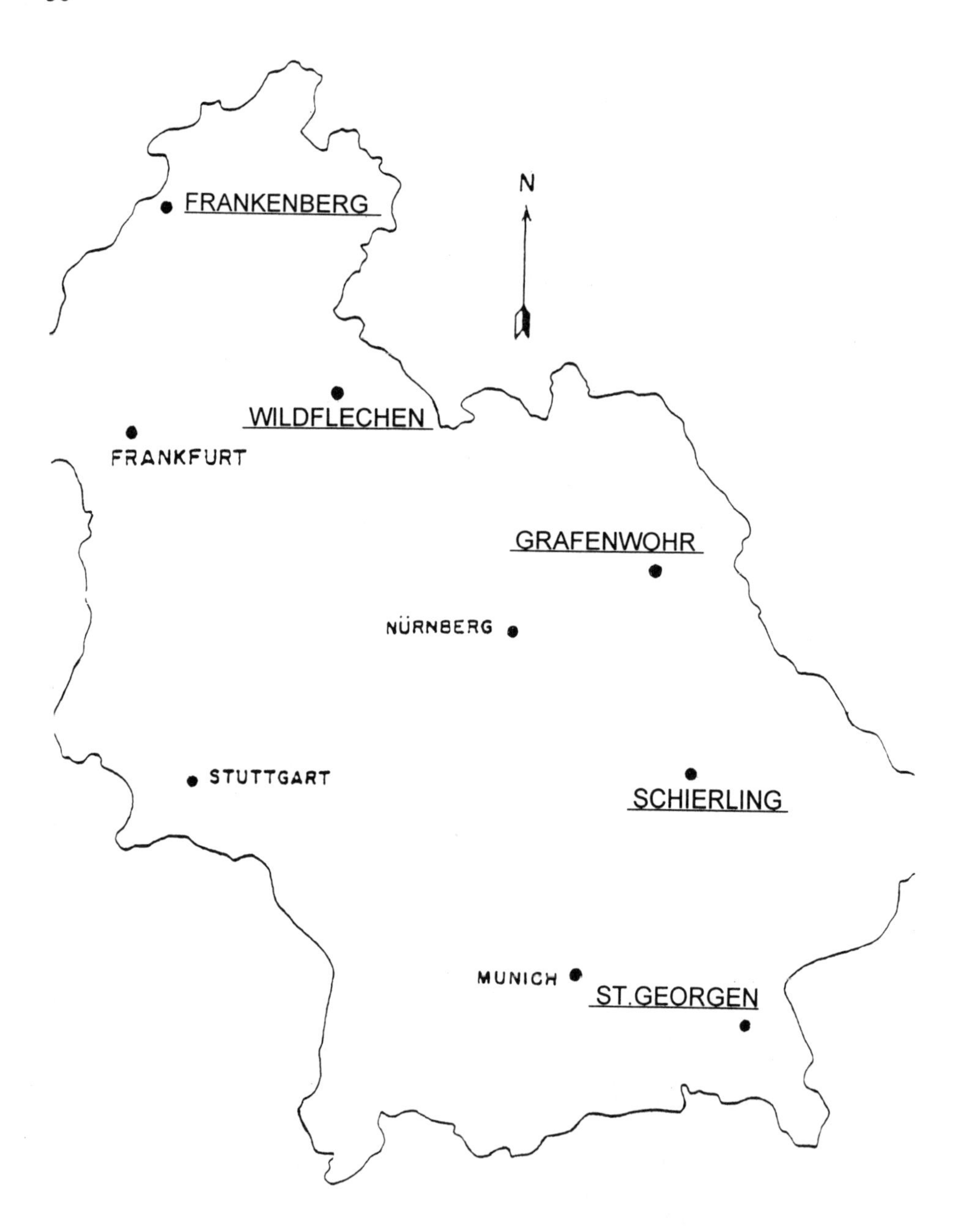

Fig. 1. US Zone of Occupation - German CML Corps Depots.

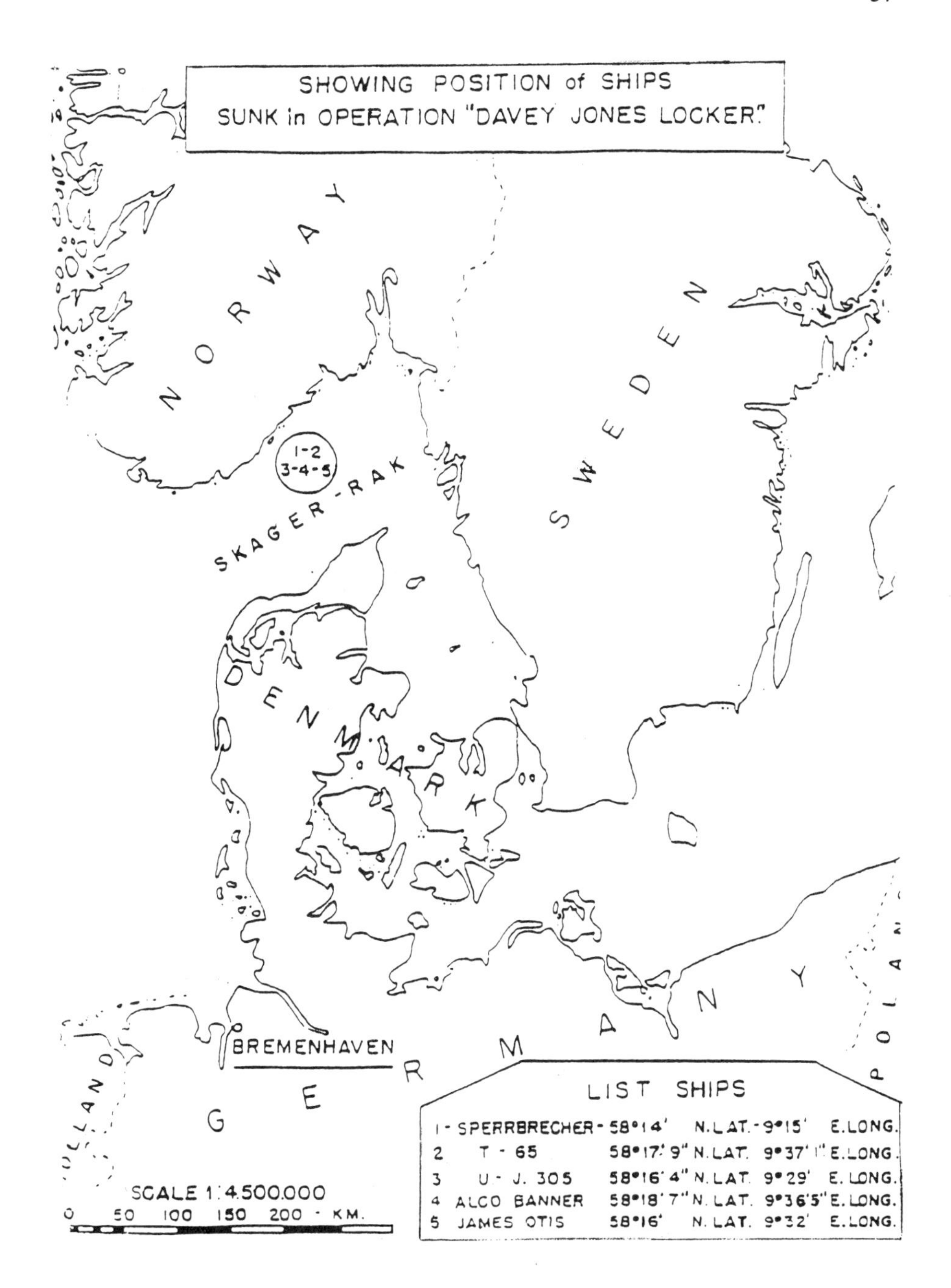

Fig. 2. Showing Positions of Ships Sunk in Operation "Davey Jones Locker".

Captured or requisitioned munitions were found in varying quantities from the following nations: Germany, U.S., England, Russia, France, Italy, Czechoslovakia, Belgium and the Netherlands. The actual outloading for all scuttling operations occurred at the Midgard Docks in Nordenham, Germany under jurisdiction of the Bremerhaven Port Chemical Officer.

The tonnage from the five depots totals anywhere between 31,000 to 39,000 tonnes. All tonnages quoted in this report are approximations. Apparent contradictions are due to the fact that additional stocks of toxic materials were discovered after 1 June 1947, and that all tonnages were of necessity rough estimations as the accurate weighing of the material demilitarised was neither practical, nor necessary. In addition, the records themselves, show varying numbers.

TABLE 1. Position of ships sunk in operation "DAVEY JONES LOCKER" - 1946

NAME/ TYPE SHIP	L. TONS	DATE SCUTTLED	DEPTH (METERS)	LAT/LONG
SPERRBRECHER Ger. Minebreaker	1349	7/01/46	650	58° 14' 9° 15'
T-65 Ger. Flak Ship	1526	7/01/46	650	58° 17' 9" 9° 37' 1"
U.-J. 305 Ger. Trawler	671	7/02/46	650	58° 16' 4" 9° 29'
ALCO BANNER	2765	7/14/46	650	58° 18' 7" 9° 36' 5"
JAMES OTIS Am. Liberty	3653	8/30/46	680	58° 16' 9° 32'

The *Sperrbrecher*, *T-65*, and *U.-J. 305* were towed out to sea by tugs. It is not known how the *Sperrbrecher* was sent to the bottom. The seacocks were opened and shells were fired into the *T-65* and *U.J.-305* hulls at the water line in the bow and stern where no toxics were stored. The *Alco Banner* took forty-five minutes to sink and was sent to the bottom by shell fire. It sunk by plunging her bow in and going under without capsizing. The *James Otis* was the last hulk to be scuttled in 1946. It is not known how it was sent to the bottom.

Figure 3 expands on the efforts of the map in Figure 2, identifying all eleven hulks, the depths, contents and chemical agents involved.

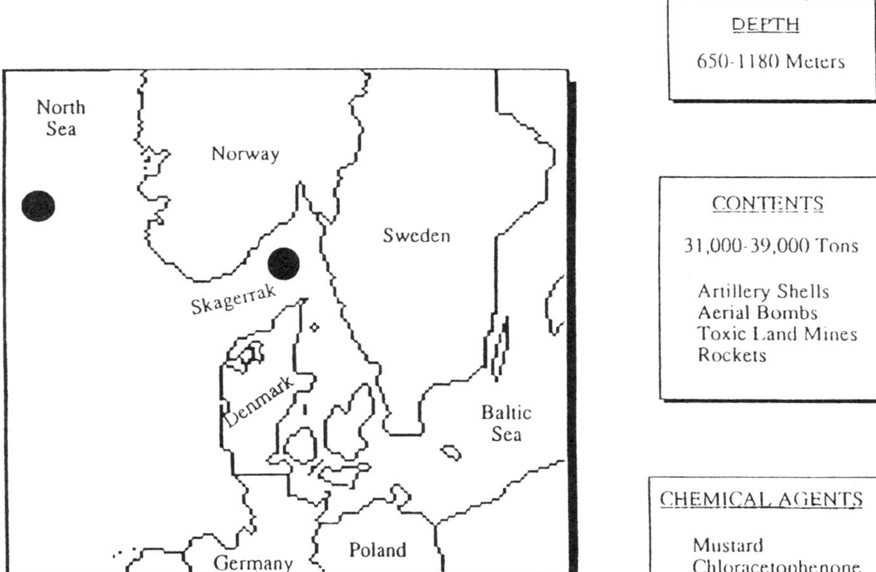

Fig. 3. Operation "Davey Jones Locker" 1946 - 1948

TABLE 2. "DAVEY JONES LOCKER", June 1947 - August 1948

NAME/ TYPE SHIP	L. TONS	DATE SCUTTLED	DEPTH (METERS)	LAT/LONG
JAMES SEWELL Am. Liberty	4000	6/6/47	725	58°15'02" 09°30'06"
JAMES HARROD Am. Liberty	3000	6/20/47	665	58°16'00" 09°33'00"
GEORGE HAWLEY Am. Liberty	1000	6/30/47	685	58°18'05" 09°38'00"
NESBITT Am. Liberty	6000	7/18/47	580	58°18'05" 09°30'00"
PHILIP HEINIKEN German Freighter	2000	7/24/48	1035	62°57'00" 01°32'00"
MARCY German Freighter	2500	8/24/48	1180	62°59'00" 01°23'00"

Final overall figures for the disposition of all toxic material captured in the U.S. Zone of Germany are shown below:

Disposition	Long Tons
Shipped to U.S.	12,000
Shipped to UK	8,000
Shipped to Italy	10,500
Sold to STEG	8,500
Scuttled at sea	31,000-39,000
Destroyed in Place	32,500
TOTAL:	102,500 - 110,500

INVESTIGATION OF A DUMPING AREA IN THE SKAGERRAK 1992

PER OLOF GRANBOM
National Defence Research Establishment
Department of NBC Defence
S-901 82 Umeå, Sweden

Keywords

Chemical ammunition, Dumping, The Skagerrak, Sediment, Analysis, Mustard Gas, Enzymeactivity, Mussels, Crabs.

Abstract

An expedition was performed in the Skagerrak in 1992 for investigation of an area, where ships with German chemical ammunition was scuttled after the Second World War. The position of the wrecks was registered. Samples of sediment taken at one of the wrecks have been analysed and proven to contain traces of mustard gas. The concentration is in the ppt-range. The enzyme activity measured in mussels and crabs placed in cages close to some wrecks showed no significant difference with those placed one km away as reference.

1. Introduction

This is a resumé of an investigation of a dumping area for chemical ammunition in the Skagerrak off the Swedish west coast. The investigation was carried out in 1992.

Dumping of German chemical ammunition was performed by UK during the years 1945-1947. The ammunition was loaded in ships, which were scuttled in the Skagerrak. The most frequently used dumping area was a 700 m deep site south-west of the town of Arendal on Norway's south coast. More than 30 ships have presumably been sunk in this area.

Another dumping area off the Swedish west coast, which is indicated in the chart (see Figure 1), has been forgotten for a long time, but journalists who have studied documents about the dumping activity called the attention to the fact that at least nine ships, among them two large cargo ships containing as much as 20,000 tons of chemical ammunition, had been scuttled in this area. The content is not known but there is reason

A. V. Kaffka (ed.), Sea-Dumped Chemical Weapons: Aspects, Problems and Solutions, 41–48.

to believe that they consist of all types of chemical weapons in the German arsenal, mainly mustard gas but presumably also the nerve gas tabun. Mass media warned about the danger this cargoes might represent when the ammunition corrodes and the chemical warfare agent is exposed to the water.

The dumping area is situated on the Swedish part of the continental shelf, therefore the National Maritime Administration was given the assignment to estimate the exact position of the wrecks and to investigate some of the sunken ships.

2. Performance of the Investigation

A survey vessel operated by the Geological Survey of Sweden surveyed the area where the scuttled ships ought to be found. A side scan sonar was used. Nineteen objects were found and 16 were believed to be wrecks or parts of wrecks. They were accurately positioned.

A group of five ships laid close together as if they were connected to each other. It was decided that this group should be investigated first.

A minelayer with a remotely controlled submarine, one-manned submersible, was chartered. It was equipped with scanning sonar, echo sounder, low-light cameras, headlamps and manipulators and conducted by cable. The submarine was brought down to the group of five wrecks and a video film was made.

The investigation was difficult to carry out due to bad visibility in the water, a strong surface current and strong currents at the seabed level, that went in different directions. The depth of 200 metres made the manoeuvring of the submarine difficult because of the effect of the currents on the umbilical cord between the minelayer and the submarine. Part of the wrecks, which were observed, were covered by trawl nets which to a high degree prevented the free movement of the submarine.

Analysis of the videotape taken from the submarine has resulted in the certain identification of the German minesweeper M16, which has been reported sunk in the area. Any grenades or bombs that could have contained chemical agents were not observed visually.

Samples of sediment were taken with a gravity corer for analysis, four samples next to the wreck and two reference samples at a distance of one kilometre from the wreck. The sediment samples were analysed chemically for the presence of mustard agent by the National Defence Research Establishment. Mustard agent has a very low water solubility. Especially viscous mustard has a very long life in water and can remain on the seabed for a very long time after that the ammunition has corroded.

3. Analysis of Sediment Samples

Layers of the sediment cores between 0-4 centimetre and 4-8 centimetre from the seabed level were taken out for analyses. Some replicate analyses were made on the same sample.

Extracts from the sediment cores were analysed in a GC/HRMS, gas chromatography/high resolution mass spectrometer. The result is demonstrated in the table in Figure 2.

Surprisingly, mustard agent in very low concentrations, exceeding the detection limit of 0.1 ppt, could be detected in some samples. It can also be seen that the mustard agent is not homogeneously distributed in the sediment. The highest concentration of mustard agent, 190 ppt, was found in a reference sample one kilometre away from the wreck. Samples 1-4 was collected next to the wrecks and 5-6 one kilometre away.

The analysis result from the sediment sample with the concentration 190 ppt is demonstrated in the chromatogram in Figure 3. The upper trace is from the molecularion of mustard agent, where both chlorine atoms have the atomic weight of 35. The middle trace is a confirming ion from the mustard agent where one of the chlorine atoms is the naturally occurring atom with the atomic weight 37. The lower trace is from the internal standard, deuterated mustard agent added to the sample.

The limited number of samples taken is not a significant statistical basis to specify which concentration of mustard agent there is in the sediments of the seabed. The analysis only show the presence of very low concentrations in sediments at a distance of at least one kilometre from the wrecks.

4. Toxicological Investigation

Possible toxic influence from the nerve gas tabun on living organisms was also investigated by placing biological samples near the wrecks. Cages with crabs and mussels were placed to leeward of the bottom current in the immediate vicinity of the five wrecks that had been the main objects of the investigation. These were compared with animals placed on a reference station one kilometre from the wrecks.

Possibly leaking nerve agent could be detected through examination of the effects on the enzyme system of the specimens.

By choosing a relatively long period of exposure (2-3 weeks) it was assumed that a single leakage at any time during the period of exposure should be detected.

By using a known technique (developed by Galgani and Bocquene) to define the enzyme acetylcholinesteras (AChE) in animal tissue, a possible influence of organic phosphorous compounds, e.g. tabun, can be registered as an inhibited enzymeactivity. The measurement is done so that the crab or mussel is homogenised and mixed with a known amount of acetylcholine (Ach, the substrate which is broken down by AChE) and a substance that forms a yellow coloured complex with broken down Ach. The more AChE in the sample the more Ach is broken down, and more yellow is the mixture. The intensity of the colour is measured with a spectrophotometer and registered as Mann and Whitney U-test, units/μg protein.

When the cages were retrieved after 17 days the specimen were quite normal and no mortality was noted. The result of the measurement is demonstrated in Figure 4, which demonstrates the average values with standard deviations. The crabs in the cages in the reference site one kilometre away from the wrecks showed a somewhat lower enzymeactivity than the crabs close to the wrecks. However the difference in enzymeactivity is not statistically significant.

5. Discussion

After this investigation was carried out in 1992, further information of these wrecks was given in 1993 from the U.K. According to it, most of the ships scuttled on this site were not loaded with chemical ammunition. Among them was the minesweeper M16, that we identified. They are in fact scuttlings of German naval craft under the requirement of the Tripartite Naval Commission. It was however confirmed that at least two large cargo ships have also been sunk in the area. Presumably viscous mustard agent is spread as small particles from these wrecks.

In a resume of the investigation we have to pay most attention to the sediment analysis, that indicated mustard agent in low concentrations. Before further conclusions can be drawn the wrecks loaded with chemical ammunition must be identified and their positions estimated. More sediment analysis must be carried out to map the spreading of the mustard agent. There are however no plans for a new expedition to this dumping area.

Literature

Report on Investigation of the Existence of Dumped Chemical Weapons on the Swedish Part of the Continental Shelf (in Swedish, summary in English), National Maritime Administration, Sweden 1992.

Fig. 1. Indication of the dumping area investigated

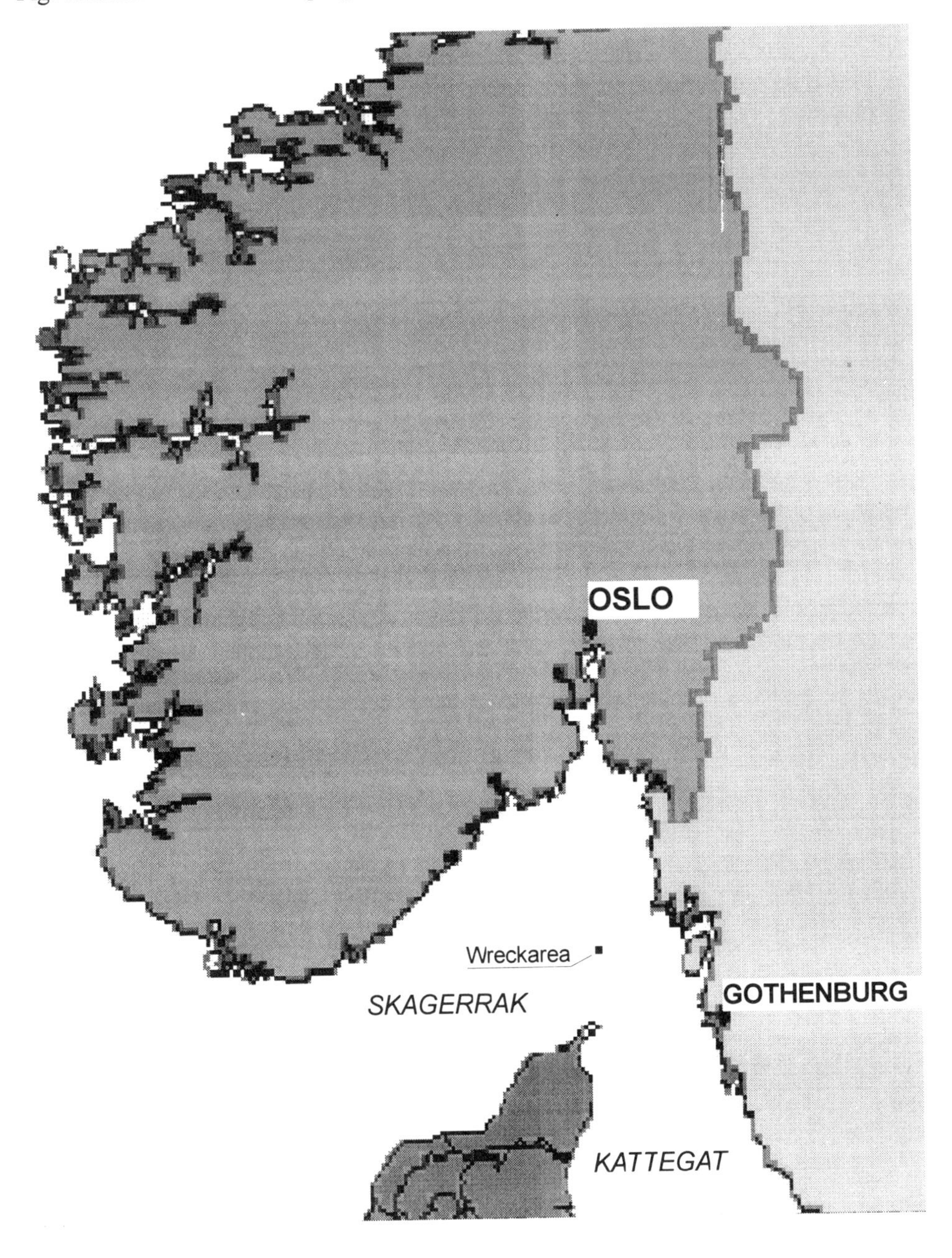

Fig. 2

Concentration (ppt = parts per trillion) of mustard agent in sediment samples collected in the Skagerrak and analysed in GC/HRMS

	Concentration (ppt)	
Sediment No.	0-4 cm	4-8 cm
1	n. d.	n. a.
2	3.2 0.1 n. d.	n. a.
3	n. d.	n. a.
4	0.2 n. d.	n. a.
5	n. d.	0.3 n. d. n. d.
6	1.2 190 n. d.	0.8 n. d. 18

n.d. not detected
n.a. not analysed

NATIONAL DEFENCE RESEARCH ESTABLISHMENT
Department of NBC Defence, Umeå, Sweden

Fig. 3

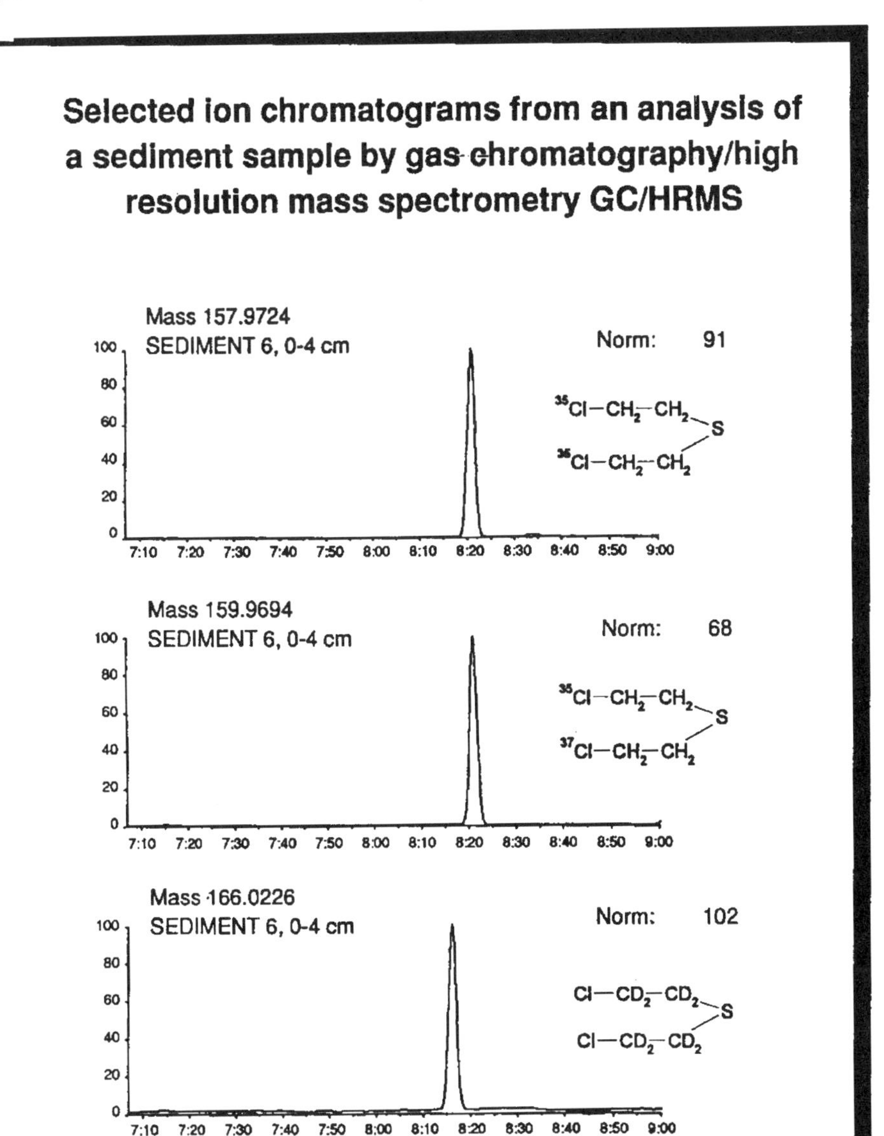

NATIONAL DEFENCE RESEARCH ESTABLISHMENT
Department of NBC Defence, Umeå, Sweden

Fig. 4

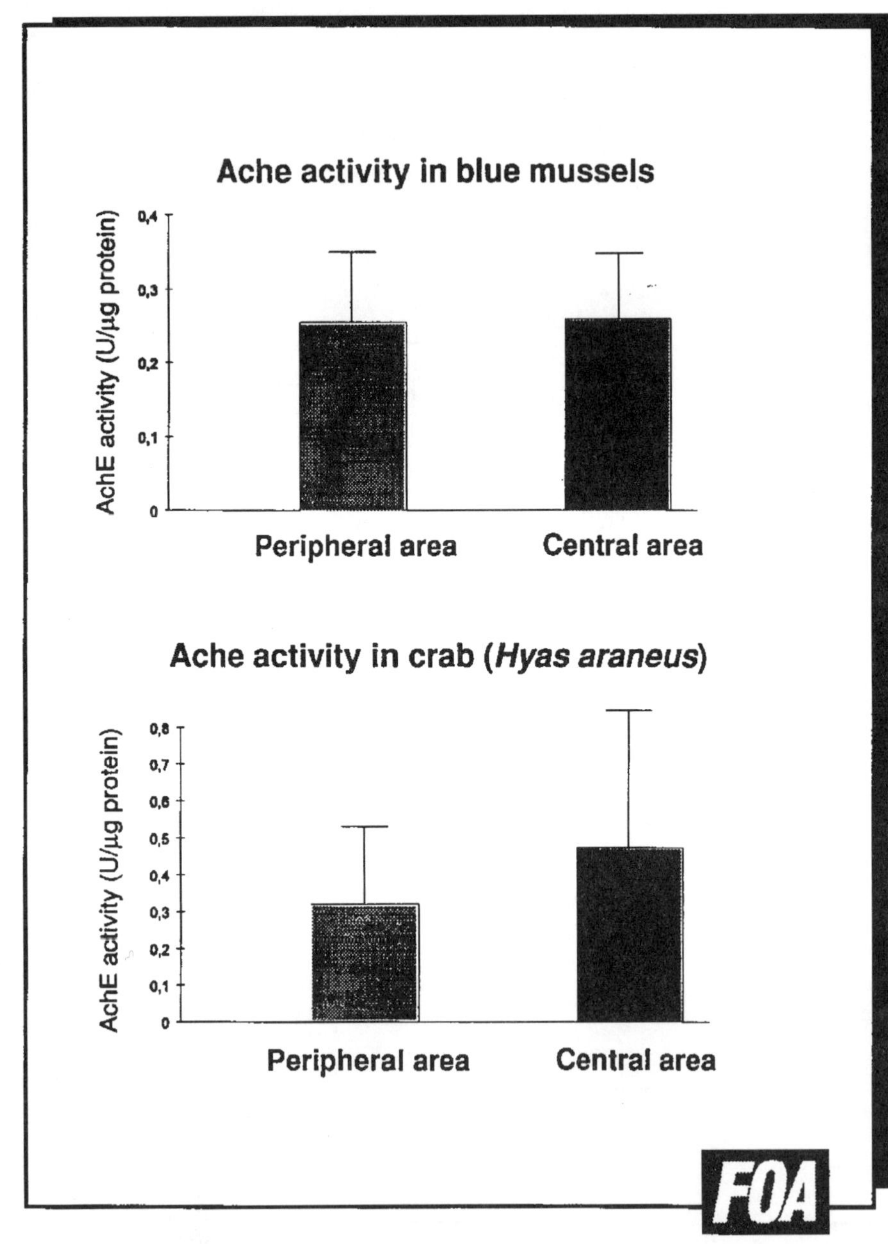

NATIONAL DEFENCE RESEARCH ESTABLISHMENT
Department of NBC Defence, Umeå, Sweden

SEA-DUMPED CHEMICAL WEAPONS AND THE CHEMICAL WEAPONS CONVENTION

DR. THOMAS STOCK
*SIPRI**
Stockholm, Sweden

1. Introduction

During the late 1980's the public became generally aware of the fact that another chemical time bomb is ticking: the chemical weapon (CW) munitions which were dumped after World War II in the Baltic and the North Sea. At the end of World War II, the Allies were faced with the problem of how to get rid of more than 300,000 tonnes of CW munitions which were left especially in Germany. The easiest and quickest solution was to load the weapons onto disabled ships and sink these ships in deep waters. Bombed and burned hulks were collected from all over northern Europe and patched up for their last journey.

Some information is available with respect to the direct effects of CW agents such as mustard gas (the major part of dumped munitions contains this agent), but only little is known about the long-term behaviour of such ammunition "stored" on the bottom of the sea.

Especially in recent years, many newspaper articles and televised reports have covered the problems related to the CW dumped into the Baltic and the North Sea, and an alarm has been sounded as to the risk of an environmental catastrophe if the chemical warfare agents are released from the corroded ammunition. Not all the information presented has been serious or truly scientifically based. It should also be remembered that in 1992 reports appeared that in the vicinity of the Danish island Bornholm, where artillery shells containing mustard gas and tabun were dumped, a large gas bubble composed of warfare agent gas had formed on the bottom of the Baltic Sea. This was soon dismissed by experts as scientifically unfeasible[1].

This paper analyses the scope of the problem of sea-dumped CW, especially in the Baltic Sea and the North Sea, and the possible consequences for the marine

---

* The views expressed in this paper are those of the author and do not necessarily reflect those of the Stockholm International Peace Research Institute (SIPRI).

[1] It is, however, possible to talk about 'rotten' gases. see: 'Giftgasblase Seifenblase?', *Frankfurter Rundschau*, 3 Mar. 1992; 'Giftgas-Bergung gefährlich', *Frankfurter Rundschau*, 5 Mar. 1992, p. 4; 'Entwarnung für Bornholm', *Frankfurter Rundschau*, 16 Mar. 1992, p. 4; 'Keine Giftgas-Blase südlich von Bornholm', *Süddeutsche Zeitung*, 16 Mar. 1992, p. 5.

A. V. Kaffka (ed.), Sea-Dumped Chemical Weapons: Aspects, Problems and Solutions, 49–66.

environment. In addition, activities of the Baltic Marine Environment Protection Commission or Helsinki Commission (HELCOM), which established in 1993 an *Ad Hoc* Working Group on Dumped Chemical Munitions (CHEMU), are briefly reviewed and recommendations for possible further action are presented. With the finalising of the Chemical Weapons Convention (CWC) in 1992 the first CW disarmament treaty was established. The treaty's treatment of sea-dumped CW and possible implications for future action if the treaty enters into force are discussed.

2. CW Dumped after World War II into the Baltic and the North Sea

After World War II, the Allied leaders faced one particular problem among many others: How would it be possible to dispose of the mostly German CW stocks as quickly and safely as possible? According to Article 3 of the Potsdam Agreement, Germany had to be demilitarised and all war material should be distributed to the Allies or destroyed. It was also decided that each occupation authority would be responsible for the munitions in its occupation zone. One of the favoured options was the disposal of these munitions by dumping at sea, as recommended by the Allied Control Commission's Standing Committee on War Material, in November 1945.[2]

Before addressing the historical calculations of what has been dumped, there is a need to estimate the scope of the problem and to identify the reliability of the sources used to do so. The following discussion focuses mainly on the European aspect of CW dumping, and especially on dumping in the Baltic Sea and the North Sea. When defining the amount of CW dumped, there is a necessity to define the time when dumping was conducted. From a pragmatic, but also historical point of view, there seems to be a need to define at least two dumping periods: post-World War II dumping and later dumping. The post-World War II dumping is related to the fact that after the end of the war the Allied forces were faced with the particular problem of how to cope with the enormous amount of CW and chemical warfare agents left by the former German Army.

There is today a more or less final picture of German CW production. However, owing to differing sources, the total amount of chemical warfare agents produced varies from 62,322 tonnes to 65,000 tonnes.[3] Germany had produced and accumulated before and during World War II the following amounts of CW:[4]

---

[2] UK, 'Report on sea dumping of chemical weapons by the United Kingdom in the Skagerrak waters post World War II' HELCOM CHEMU 2/2/5, 28 Sep. 1993, p. 3.

[3] Brauch, H. G. and Müller, R.-D., *'Chemical warfare, chemical disarmament'*, Berlin Publisher Arno Spitz, 1985, pp. 275—76; SIPRI, *'The Problem of Chemical and Biological Warfare'*, Vo. I, *'The Rise of CB Weapons'*, Almqvist & Wiksell, Stockholm, 1971; Groehler, O., *'Der Lautlose Tod'*, Rowohlt Taschenbuch Verlag GmbH, Hamburg 1989.

[4] *Chemische Kampfstoffmunition in der südlichen und westlichen Ostsee: Bestandsaufnahme, Bewertung und Empfehlungen*, Bericht der Bund/Länder-Arbeitsgruppe Chemische Kampfstoffe in der Ostsee, Bundesamt für Seeschiffahrt und Hydrographie, Hamburg, 1993.

Chloroacetophenone	7,100 tonnes
Diphenylchloroarsine	1,500 tonnes
Diphenylcyanoarsine	100 tonnes
Adamsite	3,900 tonnes
Arsin oil	7,500 tonnes
Phosgene	5,900 tonnes
S-mustard gas	25,000 tonnes
N-mustard gas	2,000 tonnes
Tabun	12,000 tonnes
Total	**65,000 tonnes**

According to historical documents[5] the following amounts of CW were discovered, destroyed, dumped or recycled on German territory between the end of World War II and the end of 1947:

In the American occupation zone	93,995	tonnes
In the British occupation zone	122,508	tonnes
In the French occupation zone	9,250	tonnes
In the Soviet occupation zone	62,505—70,500	tonnes[6]
Total	**288,258—296,203**	**tonnes**

Depending on the type of ammunition, the content of warfare agent ranges from 10 per cent (for artillery shells) to a maximum of approximately 60—70 per cent (for aerial bombs)[7] of the cross weight of the ammunition. Based upon this estimate, a rough calculation of the total amount of CW ammunition (taking the total amount of produced German chemical warfare agents of 65,000 tonnes as the point of departure) would be 260,000 to 433,000 tonnes.

---

[5] Brauch, H. G. and Müller, R.-D., *'Chemical warfare, chemical disarmament'*, Berlin Publisher Arno Spitz, 1985, pp. 275—76.

[6] The Russian report to HELCOM (Russia, 'Complex analysis of the hazard related to the captured German chemical weapon dumped in the Baltic Sea', HELCOM CHEMU 2/2/1/Rev.1, 27 Sep. 1993) cited for the first time for the Soviet occupation zone an amount of 70 500 tonnes which contained 62 500 tonnes of munition and 8000 tonnes of containerised chemical warfare agents. The 62 500 tonnes are identical to numbers presented in other reports, such as the German report to HELCOM (*Chemische Kampfstoffmunition in der südlichen und westlichen Ostsee: Bestandsaufnahme, Bewertung und Empfehlungen*, Bericht der Bund/Länder-Arbeitsgruppe Chemische Kampfstoffe in der Ostsee, (Bundesamt für Seeschiffahrt und Hydrographie: Hamburg, 1993). For the American, British and French occupation zone there are also figures presented with respect to containerised chemical warfare agents, so that the overall figures in the Russian report to HELCOM for these 3 zones are higher.

[7] Sjöfartsverket, 'Rapport om kartläggning av förekomsten av dumpade kemiska stridsmedel på den svenska delen av kontinentalsockeln', 1992, p. 13.

In the course of the HELCOM CHEMU meetings, Germany[8] and Russia[9] have provided information on dumping in the Baltic Sea, and the USA and the UK on dumping in the North Sea. The USA and the UK have both stated that their dumping activities after World War II were not performed in the Baltic Sea. That information is presented in Table 1.

TABLE 1. The Baltic Sea

Location	Quantities of munitions in tonnes	Quantities of warfare agents in tonnes	Types of warfare agents
Bornholm Basin (East) (German report)[10]	35,300——43,300	5,300——6,500	Clark II, adamsite, Clark II, adamsite, chloroacetophenone
Bornholm Basin (North-east) (Russian report)[11]	62,000[12]	11,077	Mustard gas, arsenic compounds, adamsite, chloroacetophenone, 'cyclone B'
Gotland Basin (South-east) (German report)	2,000	300	
65 miles south-west of the port of Liep (Identical south-east of Gotland) (Russian report)	5,400	958	Mustard gas, arsenic compounds, adamsite, chloroacetophenone
Bornholm Basin[13] (South-west) (German report) (Not verified)	15,000	2,250	Unknown

---

[8] *Chemische Kampfstoffmunition in der südlichen und westlichen Ostsee: Bestandsaufnahme, Bewertung und Empfehlungen*, Bericht der Bund/Länder-Arbeitsgruppe Chemische Kampfstoffe in der Ostsee, (Bundesamt fur Seeschiffahrt und Hydrographie: Hamburg, 1993).

[9] Russia, 'Complex analysis of the hazard related to the captured German chemical weapons dumped in the Baltic Sea', HELCOM CHEMU 2/2/1/Rev.1, 27 Sep. 1993.

[10] *Chemische Kampfstoffmunition in der südlichen und westlichen Ostsee: Bestandsaufnahme, Bewertung und Empfehlungen*, Bericht der Bund/Länder-Arbeitsgruppe Chemische Kampfstoffe in der Ostsee, (Bundesamt für Seeschiffahrt und Hydrographie: Hamburg, 1993).

[11] Russia, 'Complex analysis of the hazard related to the captured German chemical weapon dumped in the Baltic Sea', HELCOM CHEMU 2/2/1/Rev.1, 27 Sep. 1993.

[12] The Russian report (Russia, 'Complex analysis of the hazard related to the captured German chemical weapon dumped in the Baltic Sea', HELCOM CHEMU 2/2/1/Rev.1, 27 Sep. 1993) presents a ratio of 80% artillery munition and 15% aerial munitions. For the calculation of artillery muntion a factor of 0,15 according the warfare agent weight was used and for the aerial munition a factor of 0,6.

[13] There are witness reports that in 1946, 4 ships containing around 15 000 tonnes were dumped south-west of Rönne (Bornholm). See *Chemische Kampfstoffmunition in der sudlichen und westlichen Ostsee: Bestandsaufnahme, Bewertung und Empfehlungen*, Bericht der Bund/Lander-Arbeitsgruppe Chemische Kampfstoffe in der Ostsee, (Bundesamt für Seeschiffahrt und Hydrographie: Hamburg, 1993).

TABLE 1 *(Continued)*

Location	Quantities of munitions in tonnes	Quantities of warfare agents in tonnes	Types of warfare agents
Little Belt (German report) (Dumped by Germany shortly before the end of World War II)	5,000	750	Tabun, phosgene, chlorobenzene
Total	**Max 87,400**	**Max 15,050**	

For the area of the North Sea (Table 2), information was presented by the USA[14] and the UK[15] to the HELCOM. The main dumping area chosen by the USA during their operation 'Davey Jones Locker' (1946—48) was located in the Skagerrak, off the Swedish west coast, near Måseskär, at a depth of 650—1180 meters. The UK operation took place between 1945—47, and under British command, 34 ships containing together approximately 127,000 tonnes[16] of chemical warfare ammunition and conventional ammunition were scuttled 25 nautical miles south-east of Arendal. As regards other information on dumping operations, the picture may still be incomplete.

TABLE 2. The North Sea

Location	Quantities of munitions in tonnes	Quantities of warfare agents in tonnes	Types of warfare agents
Skagerrak (American dumping)	31,000—39,000		Mustard gas, chloroacetophenone, phosgene, tabun
Skagerrak (British dumping)	127,000		
Total	**Max. 168,000**		

If the overall amount which might have been dumped into the Baltic and the North Sea is compared to the amount of captured German CW after World War II, approximately 85 per cent has been disposed of at sea. It must be recognised that the information thus far submitted is probably not 100 per cent accurate, as the states

[14] USA, 'Study of the sea disposal of chemical munitions', HELCOM CHEMU 2/2, 8 Sep. 1993.

[15] UK, 'Report on sea dumping of chemical weapons by the United Kingdom in the Skagerrak waters post World War II' HELCOM CHEMU 2/2/5, 28 Sep. 1993.

[16] This figure includes 3 000 tonnes from the French occupation zone and up to 10 000 tonnes from the American Zone, the latter was dumped by the UK on the Allies' behalf.

involved in the dumping appear to have difficulties in retrieving the historical records completely.[17]

3. Chemical Warfare Agents and Their Behaviour in the Sea Environment

When discussing the problems of sea-dumped CW, types of chemical warfare agents related to this kind of disposal must be kept in mind. Most of thc agents are substances produced before and during World War II. According to the literature, the German chemical warfare agents produced then and mainly disposed of after the war by dumping are the following:

1. Non-lethal compounds
   Chloroacetophenone, adamsite, diphenylchloroarsine, diphenylcyanoarsine, triphenylarsine.
2. Lethal compounds
   Tabun, phosgene, mustard gas, mustard gas with additives, phenyldichloroarsine, arsinictrichloride, nitrogen mustard gas.

The behaviour of chemical warfare agents in the sea environment depends both on the chemical and the physico-chemical properties of the substances as well as external factors such as temperature, salinity, pH value and turbulence in the water. In the Baltic Sea, for instance, the pH value does not vary much—it is slightly alkaline (pH 8). Salinity and temperature are therefore the main factors which will influence the chemical reaction. Solubility and reaction speed increase if temperature increases. With a temperature increase of 10oC for example, the speed of a chemical reactions doubles.

Taking the physico-chemical properties as a point of departure, the melting and boiling points of most of the above-mentioned agents demonstrate that they are liquid or solid at 20oC. Only phosgene is a gas above 8oC. Also significant for the behaviour of chemical warfare agents is their particular vapour pressure, which is important for vaporisation of the agent.

The density of the chemical warfare agent is another factor which will decide if the agent will sink to the sea-bed or rise to the surface of the water. With the exception of tabun (density: 1.07 g/cm^3) all the agents of concern have a higher density than the Baltic sea water (density approximately: 1.08 g/cm^3). Based upon this, it can be assumed that dumped chemical warfare agents will normally not rise to the surface of the water and drift there.

For the degradation of chemical warfare agents, dissolution in water is the first and most important step. Dispersion in the sea (drifting, dilution) will occur orders of magnitude faster if the agent is already dissolved in water. The following table presents an overview of the solubility of some warfare agents.

[17] UK, 'Report on sea dumping of chemical weapons by the United Kingdom in the Skagerrak waters post World War II' HELCOM CHEMU 2/2/5, 28 Sep. 1993, pp. 78.

TABLE 3. Solubility of chemical warfare agents

Chemical agent	**Solubility in water [g l^{-1}]**
Chloroacetophenone	1
Diphenylchloroarsine (Clark I)	2
Diphenylcyanoarsine (Clark II)	2
Adamsite	negligible
Phosgene	9
Sulphur mustard gas	0.8
Nitrogen mustard gas	0.16
Lewisite	0.5
Tabun	120

As regards solubility, the reaction of chemical warfare agents with water depends on hydrolysis. This reaction leads to new compounds with properties different from those of the chemical warfare agents. In addition, they are less toxic or no longer toxic. The following discussion addresses the hydrolytic behaviour of the chemical warfare agents concerned.

Mustard gas slowly hydrolyses in water and forms hydrochloric acid and thiodiglycol. Both final products of the hydrolysis are non-toxic. The hydrolysis is dependent on temperature, density, viscosity, pH value and pressure. Because mustard gas is relatively insoluble, the slower dissolving process is the main factor of interest here. There is a huge difference between distilled water and normal sea water. In distilled water the half-life is 8.5 minutes at 25oC, while for salt water at the same temperature the half-life is 60 minutes. For sea water the hydrolysis will be slowed down by a factor of more than 3 times. The pace of hydrolysis of sulphur mustard gas also depends on the content of salt (cations and anions) in an aqueous solution. The reported half-life of sulphur mustard gas in sea-water is 15 minutes at 25oC, 49 minutes at 15oC, and 175 minutes at 5oC.[18]

With respect to the solubility of sulphur mustard gas, the following example seems illustrative. In quiescent water at 18oC, it would take 15 days for half the mass of an agent droplet with an initial diameter of 1 cm to dissolve. This is enough time for other processes to slow or even halt further dissolution and hydrolysis, such as the formation of oligomeric/polymeric layers.[19]

---

[18] Epstein, J. et al., *Summary Report on a Data Base for Predicting Consequences of Chemical Disposal Operations*, ESAP 1200-12, Headquarters, Edgewood Arsenal, Aberdeen Proving Ground, Md., 1973, quoted in *Chemical Stockpile Disposal Program: Final Programmatic Environmental Impact Statement,* vol. 3, Aberdeen Proving Ground, Md., Jan. 1988, appendix N, pp. N-15/16.

[19] Small, M. J. *Compounds Formed from the Chemical Decontamination of HD, GB, and VX and their Environmental Fate*, Technical Report 8304, US Army Medical Bioengineering Research and Development Laboratory, Medical Research and Development Command, Fort Detrick, Fredrick, Md., 1984, quoted in *Chemical Stockpile Disposal Program: Final*

Any temperature increase will speed up the hydrolysis; so for 99 per cent hydrolysis of a saturated mustard gas solution at 20oC, a time of 110 minutes is reported and at 50oC, 4 minutes.[20] Mustard gas is soluble in water with 0.8 g . l^{-1} at 20oC. Once mustard gas is dissolved, it hydrolyses quicker.

If sulphur mustard gas spilled into water bodies, the agent would sink to the bottom because its density is higher than that of water.[21] Small droplets would show some tendency to dissolve during this process, larger droplets would reach the sediment and survive there for a longer period of time. It is proven that sulphur mustard gas can resist degradation in water or in soil for years.[22]

When mustard gas is kept on the sea-bed, the degree of agitation and current velocity will have an impact on the speed of hydrolysis, too. Mustard gas at the bottom of the sea can survive in oil-like forms. An estimation of the behaviour of sulphur mustard gas indicates a theoretically long, but finite, half-life for the agent in the cold conditions of the ocean deep. Estimations have predicted that a one-tonne block of sulphur mustard gas would take about 5 years to dissolve. Because the dissolution rate is lower than the rate of hydrolysis, it will become the decisive rate.[23] Low temperature and high pressure can provide the proper circumstances for solidification of mustard gas.

The modification *Zählost* is much more stable in water. Hydrolysis in significant quantities will take place only after the diffusion of mustard gas out of the *Zählost*. The remaining thickeners will slow down further diffusion of mustard gas and will conserve the still active agent for many years. Such fragments of *Zählost* will be elastic and protected against mechanical destruction. Mustard gas which is already converted into a resin like form on its surface has a very low tendency to dissolve. Products of the initial hydrolysis form oligomeric or polymeric structures which effectively slow down the hydrolytic degradation of the remaining sulphur mustard gas. The oligomeric and dimeric products will form some kind of protective surface.

Programmatic Environmental Impact Statement, vol. 3, Aberdeen Proving Ground, Md., Jan. 1988, appendix N, p. N-16.

[20] Trapp, R., *The Detoxification and Natural Degradation of Chemical Warfare Agents*, SIPRI Chemical & Biological Warfare Studies, no. 3, SIPRI, (Taylor & Francis: London, 1985), p. 17.

[21] Robinson, J. P. Perry and Trapp, R., 'Production and chemistry of mustard gas', in J. Lundin, (ed.), *Verification of Dual-use Chemicals under the Chemical Weapons Convention: The Case of Thiodiglycol*, SIPRI Chemical & Biological Warfare Studies, no. 13, SIPRI, (Oxford University Press: 1991), p. 11.

[22] Small, M. J. *Compounds Formed from the Chemical Decontamination of HD, GB, and VX and their Environmental Fate*, Technical Report 8304, (US Army Medical Bioengineering Research and Development Laboratory, Medical Research and Development Command, Fort Detrick: Fredrick, Md., 1984), quoted in *Chemical Stockpile Disposal Program: Final Programmatic Environmental Impact Statement,* vol. 3, (Aberdeen Proving Ground, Md., Jan. 1988), appendix N, p. N-16.

[23] Sage, G. W. and Howard, P. H., '*Environmental Fate Assessments of Chemical Agents VX & HD*', CRDEC-eccr-043, Aug 89, ADB137325L, quoted in Donnelly, P., J., '*Environmental Fate of Chemical Agents in Deep Seawater: Literature Review*', 3 Aug. 1990.

Nitrogen mustards, in general, react slower than the sulphur mustards. The hydrolysis of nitrogen mustards was extensively studied recently.[24] The products of hydrolysis are non-toxic. The solubility in water for HN-3 is 0.16 g . l^{-1}. The time frame for a total decomposition in fresh water is approximately 3 weeks.[25] For the tertiary amine, HN-1, the solubility in water is much higher.

Diphenylchloroarsine (DA, Clark I) hydrolysis in water will lead to diphenylarsenious acid and hydrochloric acid and *diphenylcyanoarsine* (DC, Clark II) will lead to hydrogen cyanide and diphenylarsenious oxide. Both acids are toxic, nevertheless they will be detoxified quickly in water. Both arseno-organic compounds from the hydrolysis can later be decomposed into inorganic arsenic compounds, and they will remain toxic. As for all arsenic compounds, the possible bio-accumulation would be a problem of environmental concern.

Adamsite is more or less insoluble in water. The agent hydrolyses very slowly in water. The reaction products are hydrochloric acid and diphenylarsenious acid.

Lewisite (α-lewisite or 2-chlorovinyl dichloroarsine) is only slightly soluble in water[26] but when dissolved it hydrolyses rapidly into 2-chlorovinylarsine oxide. The oxide is fairly soluble in sea water. In an alkali solution a further reaction into acetylene and arsenic acid might be possible. With high turbulence, lewisite can dissolve more rapidly in large volumes of water.[27] Because lewisite is 36 per cent arsenic by weight, even if broken down, a toxic element would always remain.

The primary factor for the hydrolysis of nerve agents is their aqueous solubility. *Sarin* is more soluble than *soman* or *tabun*.

Tabun has a good solubility in water. The hydrolysis rate is dependent on the pH value and also the temperature, but is always rapid. The products are phosphoric acid and hydrogen cyanide. The half-life for tabun in sea water at a temperature of 7oC is approximately 5 hours.[28] It is important to note that there will always be a difference in the half-life for phosphor organic agents in fresh water compared to sea water.

Phosgene is a very reactive compound which in sea-water hydrolyses into carbon dioxide and hydrochloric acid. The reaction is very fast. If the temperature is 0oC, a solution containing 1 per cent of phosgene will be hydrolysed in less then 20 seconds.

The Table 4 presents a compilation of data concerning the half-life of some chemical warfare agents in water at different temperatures.

---

[24] *'Analytik und Reaktionsverhalten von N-Lost aus alten Kampfstoffgranaten'*, Institut für Chemie und Technologie von Erdölprodukten, Technische Universität Wien, 27 June 1990.

[25] *'Chemische Kampfstoffmuntion in der südlichen und westlichen Ostsee: Bestandsaufnahme, Bewertung und Empfehlungen'*, Bericht der Bund/Länder-Arbeitsgruppe *'Chemische Kampfstoffe in der Ostsee'*, Bundesamt für Seeschiffahrt und Hydrographie Hamburg, 1 May 1993, p. 22.

[26] *Chemical Stockpile Disposal Program: Final Programmatic Environmental Impact Statement*, vol. 3, Aberdeen Proving Ground, Md., Jan. 1988, appendix N, p. N-16.

[27] *Chemical Stockpile Disposal Program: Final Programmatic Environmental Impact Statement*, vol. 3, Aberdeen Proving Ground, Md., Jan. 1988, appendix N, p. N-16.

[28] *'Chemische Kampfstoffmuntion in der südlichen und westlichen Ostsee: Bestandsaufnahme, Bewertung und Empfehlungen'*, Bericht der Bund/Länder-Arbeitsgruppe *'Chemische Kampfstoffe in der Ostsee'*, Bundesamt für Seeschiffahrt und Hydrographie Hamburg, 1 May 1993, p. 22.

TABLE 4. Stability of chemical warfare agents in fresh water[29]

Agent	Half—life	pH	Temp. (⁰C)	Source
GA (tabun)	7 hr	4—5	20	b
	8.5 hr	7.0	20	b
GB (sarin)	(5.4x108)x10-pH hr e.g., 5.4.hr [sic]	7.0	25	a
	837 hr	7.5	0	b
	453 hr	7.5	4	b
	46 hr	7.5	20	b
	24 hr	7.5	25	b
GD (soman)	82 hr	7.0	20	a
	996 hr	7.0	25	b
	428 hr	7.5	25	b
	13 years	7.6	4	b
	184 hr	8.0	25	b
HD (sulphur mustard gas)	110 min = 99% hydrolysis		20	a
	8.5 min (distilled H_2O)		25	b
	7.4-15.8 min		20	b
	158 min		0.6	b
HN_3 (nitrogen mustard gas)	Hydrolyses slowly in water. pH—dependent quantitative hydrolysis occurs only when temperature is above 90⁰C			a
L (lewisite)	'Rapid' for dissolved L			b

a Trapp, R., *The Detoxification and Natural Degradation of Chemical Warfare Agents*, SIPRI Chemical & Biological Warfare Studies, no. 3, SIPRI, (Taylor & Francis: London, 1985).
b USA, Department of the Army, Program Executive Officer, Program Manager for Chemical Demilitarization, *Chemical Stockpile Disposal Program: Final Programmatic Environmental Impact Statement*, vol. 3, (Aberdeen Proving Ground, Md., Jan. 1988), App. N.

4. Marine Environment and Chemical Warfare Agents

Scientific knowledge about the toxic effects of chemical warfare agents on the marine environment is still limited. The main focus of study has always been on the effects on human beings, animals and the environment of the use of CW.

---

[29] Leffingwell, S. S., '*Public Health Aspects of Chemical Warfare Agents*', in Somani, S. M. (ed.), 'Chemical Warfare Agents', Academic Press Inc., San Diego, 1992, p. 329.

When considering the risk which chemical warfare agents from sea-dumped CW pose to the marine environment, there is first, of course, the step where the agent is released into the water. This implies that the munition shells have corroded to such an extent that the agent can enter the water. Depending on the rate of dissolution, solubility, rate of degradation and current conditions, different concentration levels of warfare agent are possible. For a short time high concentrations are possible and if organisms are in the immediate vicinity they can be damaged. This is, of course, a more theoretical approach owing to the complex system of factors involved in such an exposure scenario.

The corrosion of ammunition shells has been discussed in a Russian report.[30] According to this report the salinity of sea water, its temperature, ambient pressure, structure of metal and joints have no significant influence on corrosion, whereas the speed of ambient stream seems to be vitally important. Based upon mathematical calculations, the maximal rate of corrosion was predicted for sunk chemical ammunition. According to these calculations, the highest rate of release of mustard gas will be at about 125 years after dumping. However, it is necessary to note that such theoretical calculations on corrosion cannot easily be transferred into practical risk assessments.[31] Sea water conditions are more complex and several external factors could create a different scenario. Practical investigations have shown that part of the CW is already corroded.

Most information on ecotoxicity available is with respect to mustard gas and nitrogen mustard gas.[32] Tests have been performed in an aquarium where fish were exposed to different concentrations of mustard gas. Based upon the understanding that different organisms react with varying degrees of sensitivity, the following acute toxic concentrations have been estimated: for some algae around 1 mg/l and for eels and flatfish well below 10 mg/l. Test organisms have been exposed to mustard gas dissolved in water or supplied as lumps of viscous mustard gas on the bottom of a test aquarium.[33] It has been evaluated that the mustard gas has no significant effect on fish, and that fish probably do not bioaccumulate the agent.

For chemical warfare agents containing arsenic there is an additional aspect to observe. After these compounds have been degraded, inorganic arsenic compounds will still remain. Such compounds remain carcinogenic to humans, and because of possible bioaccumulation in fish they continue to pose a risk. However, the extent to which

---

[30] Russia, 'Complex analysis of the hazard related to the captured German chemical weapons dumped in the Baltic Sea', HELCOM CHEMU 2/2/1/Rev.1, 27 Sep. 1993.

[31] Granbom, P. O. , 'Dumped chemical ammunition in the Baltic: a rejoinder', *Security Dialogue*, vol. 25, no.1, (1994), pp. 105—110.

[32] Miljöstyrelsens Havforureningslaboratorium, 'Rapport om forsog over optagelse af giftgas i fisk', Jan. 1986; Fate and effects of dumped chemical warfare agents', submitted by Denmark, HELCOM CHEMU 1/5, 7 Apr. 1993; *'Chemische Kampfstoffmuntion in der sudlichen und westlichen Ostsee: Bestandsaufnahme, Bewertung und Empfehlungen'*, Bericht der Bund/Lander-Arbeitsgruppe *'Chemische Kampfstoffe in der Ostsee'*, Bundesamt fur Seeschiffahrt und Hydrographie Hamburg, 1 May 1993, p. 22; Somani, S. M. (ed.), *Chemical Warfare Agents*, Academic Press Inc., San Diego, 1992.

[33] *Fate and effects of dumped chemical warfare agents*, submitted by Denmark, HELCOM CHEMU 1/5, 7 Apr. 1993.

inorganic arsenic compounds undergo further degradation reaction in fish to form non-toxic organic arsenic compounds is still not known.

The maximum solubility of chemical warfare agents in water is around 1000 mg/l or below. Under real sea-water conditions and in an open aquatic system like the Baltic Sea, with large dilution capacity, a maximum concentration of less than 10 per cent of the theoretical solubility can be assumed for short periods.[34] Owing to further dilution processes and on-going degradation reactions, the possibility that a high concentration of chemical warfare agent will be stable over a long period of time in sea water is quite remote. For phosgene and tabun, which are readily soluble in water, the initial concentration after release may be much higher, but both substances are easily degraded and their concentrations will fall below toxicity limits within a very short time.

In summary, the available information on the poisoning of the marine environment and fish by released chemical warfare agents is very limited and mostly related to laboratory investigations. The complex ecosystem of the Baltic Sea has been studied at length, however, most investigations have been performed regarding normal environmental contaminates. There is no real expertise available on the bioaccumulation of such agents in marine organisms. Assessing the potential threat to the marine environment after the massive release of chemical warfare agents from dumped CW requires more detailed scientific investigations.

5. The CWC and Sea-dumped CW

The Chemical Weapons Convention, which was concluded in Geneva in September 1992 after more than a decade of negotiations, is a major achievement in the history of arms control and disarmament. For the first time a whole class of weapons of mass destruction, chemical weapons, will be prohibited. The provisions of the CWC include the total destruction of all existing CW and detailed verification procedures to make sure that in addition to destruction, no new CW are under development. Of major concern to the CWC are stockpiled CW, which by definition are CW produced and stockpiled since 1946. For CW before 1946 the CWC defines a category of "old chemical weapons" (OCW). The treatment of these weapons is different compared to the treatment of CW produced and stockpiled after 1946. After detailed investigation by the inspectors from the Organisation for the Prohibition of Chemical Weapons (OPCW), an evaluation of the "usability" status has to be performed which will lead to an assessment of the risk of those weapons. Following that, the procedure for their destruction has to be determined. The main questions in the context of sea-dumped CW is how the CWC will cope with these very special munitions, and whether the CWC is capable of addressing the current problems in this respect.

---

[34] *Chemische Kampfstoffmunition in der südlichen und westlichen Ostsee: Bestandsaufnahme, Bewertung und Empfehlungen*, Bericht der Bund/Länder-Arbeitsgruppe Chemische Kampfstoffe in der Ostsee, (Bundesamt für Seeschiffahrt und Hydrographie: Hamburg, 1993).

The Chemical Weapons Convention contains provisions relevant for sea-dumped CW in Article III and IV. However, it may be useful to first note the definition of CW under Article II. Under this Article, CW are defined in paragraph 1 as the following:

"Chemical Weapons" means the following, together or separately:

(a) Toxic chemicals and their precursors, except where intended for purposes not prohibited under this Convention, as long as the types and quantities are consistent with such purposes;
(b) Munitions and devices, specifically designed to cause death or other harm through the toxic properties of those toxic chemicals specified in subparagraph (a), which would be released as a result of the employment of such munitions and devices;
(c) Any equipment specifically designed for use directly in connection with the employment of munitions and devices specified in subparagraph (b).

The toxic chemical in paragraph 1 (a) is defined as:

Any chemical which through its chemical action on life processes can cause death, temporary incapacitation or permanent harm to humans or animals. This includes all such chemicals, regardless of their origin or of their method of production, and regardless of whether they are produced in facilities, in munitions or elsewhere.

From the above definition it is clear that not only the toxic chemical (chemical warfare agent) is part of the CW, but the munition and devices specifically designed for use as CW and related equipment are also covered by the definition.

Articles III and IV contain provisions which will exempt certain categories of CW from the general definition of CW. This is already done under the Article II's definitions of old and abandoned CW, two categories which will be treated differently with respect to declaration and destruction. The two additional exceptions under Article III (Declarations) and Article IV (Chemical Weapons) are related to CW which have been and remain land-disposed and CW which have been sea-dumped. The exempted CW, sea-disposed or land disposed CW, have to have been disposed of before a specific date and have to remain buried. The Article III exemption from the declaration provisions for CW reads as follows under paragraph 2:

The provisions of this Article and the relevant provisions of Part IV of the Verification Annex shall not, at the discretion of a State Party, apply to chemical weapons buried on its territory before 1 January 1977 and which remain buried, or which had been dumped at sea before 1 January 1985.

This paragraph exempts all CW which were 'sea-disposed' or 'land-buried' from all declarations under Article III and from the follow-up obligations under Part IV of the Verification Annex, which deals with the detailed verification and destruction requirements, as long as they remain buried.

It is important to note the meaning of 'at the discretion of a State Party'. If a State Party declares sea-dumping or land-burial of CW, such a declaration would be a voluntary measure, and not obligatory.

In terms of prohibiting CW and their use, which is the major purpose of the CWC, any recovery of CW from dumping sites and their respective destruction within 10 years (as required under the Verification Annex, Part IV (A)) would add nothing to the security of State Parties. The necessity of such operations in terms of ecological security is a different matter, and this was intentionally kept outside the scope of the Convention.[35]

The exemption from declarations under Article III for sea-dumped CW cannot be explained. It might be true that a detailed declaration could be difficult for some of the operations after World War II, however, on other declaration obligations, such as the declarations of past transfers of CW, the Convention requires such information to the extent possible. The main question is: Why would states which have possessed CW in the past and later sea-dumped the munitions be unable to declare such activities?

The most unclear element here is the cut-off date for the exemption, 1 January 1985. This date was included in the text at the very last moment,[36] after private consultations and without ever being publicly explained. One possible explanation is that there was an understanding to find a cut-off date which would not cause any problems for a potential State Party which had dumped CW in the past. This date 1 January 1985 leads to the thought that there might be at least one State Party which might be affected by an earlier date, or in other words, there must have been sea-dumping of CW which was conducted after the post-World War II dumping.[37]

It should be noted here that with exempting the sea-dumped CW from declaration under Article III, there was also a need to exempt them from the destruction obligations under Article IV. The text under paragraph 17 reads as follows:

The provisions of this Article and the relevant provisions of Part IV of the Verification Annex shall not, at the discretion of a State Party, apply to chemical weapons buried on its territory before 1 January 1977 and which remain buried, or which had been dumped at sea before 1 January 1985.

6. The European approach

The sea-dumped CW in the Baltic Sea and the North Sea is at first glance a European problem. The states around these seas are most concerned about possible danger to the

[35] Krutzsch, W. and Trapp, R., *'A Commentary on the Chemical Weapons Convention'*, Martinus Nijhoff Publishers, Dordrecht, Boston, London, 1994, p. 58.

[36] Krutzsch, W. and Trapp, R., *'A Commentary on the Chemical Weapons Convention'*, Martinus Nijhoff Publishers, Dordrecht, Boston, London, 1994, p. 58.

[37] At least for the USA the sea-dumping of CW is public with respect to Congress Hearings in the 60s and 70s.

marine environment and to fishing.[38] For this reason under the umbrella of the Baltic Marine Environment Protection Commission, also known as Helsinki Commission (HELCOM), a decision was made to establish an *Ad Hoc* Working Group on Dumped Chemical Munition (CHEMU) in 1993. The group held its first meeting in April 1993 in St. Petersburg.[39] The mandate of the group was defined as follows:[40]

(1) Reports by the Contracting Parties on available and correct information on dumped chemical munition, which may include: (a) location (co-ordinates) of dumping sites in the responsibility zones of the respective countries as well as the places of discovery; (b) time of dumping, type and quantity of dumped chemical munition; (c) characteristics of the dumped material (type of chemicals, etc.); (d) dumping techniques.

(2) Preparation of general information including maps on the basis of the national reports.

(3). Assessment of the possible relocation of dumped material by various means and, if possible, provision of a forecast for future locations.

(4) Assessments of the effects/hazards to the marine environment, living resources and human activities, identification of gaps of knowledge and proposals for action with a view to filling such gaps.

(5) Elaboration of a proposal for a HELCOM strategy for dumped chemical munition (draft joint action programme, guidelines, recommendations).

The HELCOM CHEMU received national reports from all the contracting parties, including reports from those countries responsible for dumping CW after World War II (Russia, the UK and the USA).

In January 1994 the Third Meeting of the *Ad Hoc* Working Group on Dumped Chemical Munition (HELCOM CHEMU) of the Helsinki Commission was held in Copenhagen.[41] During this meeting the Russian delegation stated that "the data submitted to the Ministry of Environment Protection does not contain any reference to dumping of chemical weapons in the Baltic Sea after 1947". However, there still seems to be much scepticism about this statement, especially in light of the many allegations regarding later dumping operations by the former Soviet Union, also into the Baltic

---

[38] About the danger these sea-dumped CW poses to fisheries there are several reports and papers published during the last few years. This aspect will be not discussed in this paper here. See Krohn, A. W. 'The challenge of dumped chemical ammunition in the Baltic Sea', *Security Dialogue*, vol. 25, no.1, (1994,) pp. 93—103.

[39] Baltic Marine Environment Protection Commission, Helsinki Commission, *Ad Hoc* Working Group on Dumped Chemical Munition (HELCOM CHEMU), Report of the 1st Meeting, St Petersburg, Russia, 19—21 Apr. 1993, HELCOM CHEMU 1/8. The meeting was attended by delegations from Denmark, Finland, Germany, Lithuania, Poland, Russia and Sweden and by observers from Latvia, Norway, the UK and the USA as well as from the Coalition for a Clean Baltic (CCB) and Greenpeace International.

[40] Granbom, P. O. , 'Dumped chemical ammunition in the Baltic: a rejoinder', *Security Dialogue*, vol. 25, no.1, (1994,) pp. 105—110.

[41] Press Release, Helsinki Commission, Baltic Marine Environment Protection Commission, Copenhagen, 21 Jan. 1994.

Sea.[42] The working group decided to forward to the March 1994 Helsinki Commission meeting on the ministerial level a final report[43] which included conclusions and recommendations for further actions. The report contained the recommendation "not to recover chemical munitions from the Helsinki Convention Area", because the risks connected with recovery are high. The meeting decided to prolong the mandate of the *Ad Hoc* Working Group.[44] Denmark will continue to lead the work of the group. The next two meetings took place in June[45] and September.[46] The discussion focused on: (a) investigation of the chemical processes of warfare agents and ecological effects of such processes; (b) investigations of the state of corrosion of dumped chemical munitions; (c) elaboration of the Baltic Guidelines for fishermen on how to deal with dumped chemical munitions; and (d) elaboration of the Baltic Guidelines on how the appropriate authorities should deal with incidents where chemical munitions are caught by fishermen.[47] For the last two issues, draft guidelines were developed. A report submitted earlier by Latvia stated that there have been no dumpings of chemical munitions by Latvia after the re-establishing of its independence and no further information on these issues has been obtained from Russia.[48] Also Poland informed that no dumping of chemical munitions had been carried out.[49] Germany hosted the third meeting of the *Ad Hoc* Group in December 1994. All participating states are still requested to provide any information on dumping activities, especially those after 1947.

The work of HELCOM CHEMU has been proved to be important. It is one European political forum which has been successfully established to discuss not only on a political level the specific aspects of sea-dumped CW.

7. Conclusions

The problem of CW sea-dumped after World War II in the Skagerrak and the Baltic Sea is receiving more and more public interest. Many warnings have been sounded during

---

[42] SIPRI, *SIPRI Yearbook 1993: World Armaments and Disarmament* (Oxford University Press: Oxford 1993), pp. 282-3; SIPRI, *SIPRI Yearbook 1992: World Armaments and Disarmament* (Oxford University Press: Oxford 1992), p. 172.

[43] 'Report on Chemical Munitions Dumped in the Baltic Sea', Report to the 16th Meeting of Helsinki Commission, 8-11 March 1994, from the *Ad Hoc* Working Group on Dumped Chemical Munition (HELCOM CHEMU), January 1994.

[44] Convention on the Protection of the Marine Environment of the Baltic Sea Area, 1974 (Helsinki Convention), Helsinki Commission, Baltic Marine Environment Protection Commission, Report of the 15th Meeting, Helsinki, Finland, 8-11 March 1994, HELCOM 15/18.

[45] *Ad Hoc* Working Group Dumped Chemical Munition (HELCOOM CHEMU), Report of the Fourth Meeting, Copenhagen, Denmark, 16-17 June 1994, HELCOM CHEMU 4/5.

[46] *Ad Hoc* Working Group Dumped Chemical Munition (HELCOOM CHEMU), Report of the Fifth Meeting, Copenhagen, Denmark, 22 September 1994, HELCOM CHEMU 5/4.

[47] Krohn, A. W. 'The challenge of dumped chemical ammunition in the Baltic Sea', *Security Dialogue*, vol. 25, no.1, (1994,) pp. 93—103.

[48] *Ad Hoc* Working Group Dumped Chemical Munition (HELCOOM CHEMU), Report of the Fourth Meeting, Copenhagen, Denmark, 16-17 June 1994, HELCOM CHEMU 4/5p. 4.

[49] *Ad Hoc* Working Group Dumped Chemical Munition (HELCOOM CHEMU), Report of the Fifth Meeting, Copenhagen, Denmark, 22 September 1994, HELCOM CHEMU 5/4, p. 4.

recent years that these CW might lead to an environmental catastrophe and there is a need to take action as soon as possible. However, the investigations conducted in the course of HELCOM CHEMU and the few scientific based publications from the last 3 years have made the whole discussion more issue-focused and substantial. This paper has attempted to analyse the problem and show ways and directions where action might be possible. The main conclusions are the following:

1. The general locations of CW which have been dumped into the Baltic Sea and the North Sea after World War II are known as is their total amount and the agents contained in them. However, there is a need to conduct further searches in historical archives, and additional research is needed to complete the picture of what has been dumped.

2. There is a need to conduct more comprehensive investigations, with the technical means available now, to identify the dumping areas as precisely as possible.

3. There is a need to inform the directly concerned community (fishermen) more comprehensively to avoid any threat which might be caused by recovery of individual munitions. Fishermen have to be educated about the proper procedures if CW are recovered.

4. Any chemical warfare agent released into sea water will undergo degradation, which will result in less toxic or non-toxic agents. However, there is a need to investigate the degradation of the agents concerned under actual conditions, because the 'sea' ecosystem is not comparable with the laboratory environment.

5. There is a pertinent need to conduct a more detailed scientific investigation of the bioaccumulation of chemical warfare agents in fish and other marine animals. There is a necessity to conduct risk-assessments under actual conditions.

6. Sea-dumped CW should be examined for corrosion to predict more realistically the impact of the sea environment on munition shells and to forecast possible major releases of chemical warfare agents in the future.

7. For each dumping area there is a need to conduct a specific risk assessment, including recommendations for further action.

8. Pilot studies are recommended to assess the feasibility (in technical and operational terms) of CW recovery. When CW are recovered the capacity must be available to destroy them immediately afterwards. Storage of recovered CW is not a practical solution because it only creates new problems.

9. The CWC does not provide the legal basis to recover CW which where sea-dumped before 1985. Any declaration by a treaty party is only voluntary. As long as these CW remain sea-dumped, there is no obligation to destroy them. Only after recovery might these munitions fall under the category of "old CW" and have to be destroyed under the definite time-limit and under monitoring by international inspectors from the OPCW. If sea-dumped CW are recovered in large amounts, additional costs and manpower (inspectors) will be required in the OPCW.

10. The HELCOM CHEMU should continue its work with the aim of providing not only a discussion forum for the concerned states, but also to promote co-operation between the concerned states.

11. It will be extremely expensive to investigate, recover and subsequently destroy sea-dumped CW, and there is an excellent opportunity for relevant industries to assist in such operations.

Last, but not least, sea-dumped CW may pose a general risk to the environment. If this is the general perception, a real possibility exists to take further action in the context of European co-operation ranging from detailed investigation of the status of these munitions to possible recovery and destruction.

HOW TO SAVE THE BALTICS FROM ECOLOGICAL DISASTER

GENERAL LEUTENANT (RET) B.T. SURIKOV
Institute of the USA and Canada Studies
Russian Academy of Sciences, Moscow, Russia

By the time World War II broke out the USA had 135,000 tonnes, and the UK had 35,000 tonnes of different types of chemical weapons (CW). The Allies developed detailed strategic plans for the use of massive amounts of chemical weapons against Germany. During the World War, Wehrmacht kept in good fighting order its chemical forces -- military units able to use many different CW. In 1937 the General Staff of German Land Forces determined a strategy for the mass use of chemical weapons: "We must not make the mistakes of the First World War once again by using the new CW in small, uncoordinated amounts. These CW ought to be used with lighting speed, unexpectedly, precisely and at the wide front" [1].

Wehrmacht was prepared for use of massive amounts of CW for a period of 5 months. Chemical forces were armed with bombs, shells, mines of diverse calibres, high-explosive bombs, grenades and poison smoke-pots. In addition, Wehrmacht was well-equipped with special devices for fast contamination of the ground with stable toxic substances. Germany had a large stock of yperite, lewisite, adamsite, phosgene, diphosgene and chloracetophenol.

By the beginning of World War II, the Soviet Union also had a large stock of all types of poison substances excluding the organophosphorous ones. Documents kept in military archives say that over the period of the war Soviet military and political leaders let chemical forces undertake only chemical reconnaissance and decontamination of arms, uniforms and grounds. In addition, chemical forces could use flame-throwers and camouflage smokes.

The Soviet secret service revealed in advance Hitler's intention to use the poison gases on the Eastern Front in 1942. This most important dispatch was delivered to Moscow from Berlin in the spring of 1942. At that time the Soviet Union warned Germany through diplomatic channels in one of the neutral countries that it would undertake adequate steps in response. This information also reached the coalition Allies. On June 5, 1942, President Roosevelt warned Hitler on behalf of the United States and Great Britain that in the case of chemical warfare against either of the anti-Hitler coalition countries, these countries would act the same way. Fortunately, mankind escaped large-scale chemical warfare at that time.

However, the present generation inherited a dangerous legacy from the past -- lethal chemical weapons not used during World War II.

A. V. Kaffka (ed.), Sea-Dumped Chemical Weapons: Aspects, Problems and Solutions, 67–70.

The death of huge masses of starfish and crabs in May 1990 in the White Sea, caused possibly by sea-disposed yperite, reminded me of the prominent Soviet naval commander Admiral V.F. Tributs, who shared with me his fears concerning CW dumped in the European seas. During the period of 1939-1947, Admiral Tributs was the commander of the Baltic Fleet, and was well-informed about the secret operations carried out by the Soviet Union, the USA, and the UK aimed at the quickest elimination of 260,000 tonnes of chemical weapons found in Germany. Tributs told me about CW disposed of in the Baltic sea and neighbouring channels as a potential threat to future generations. He also repeatedly stressed that, at that time, neither victors nor defeated had adequate technologies to safely destroy CW in their armouries. This was the reason why political leaders of the Soviet Union, the USA and the UK made the fundamental decision to bury CW in the deeps of the Atlantic Ocean. Unfortunately, though worthwhile from the ecological point of view, this idea could not be put into effect for technical reasons. That is why the European seas are now filled with a huge amount of CW which are extremely dangerous for the millions of people who live in this region.

In 1945 the Soviet troops discovered only 35,000 tonnes of Wehrmacht's chemical ammunition in the Soviet occupation zone, while the rest 215,000 tonnes were kept in the Western zones. Fortunately, our military archives keep 52 well-saved pages of classified documents regarding the German CW scuttled in 1947 in the Baltic Sea by the Soviet troops. My former colleagues from the Defence Ministry broke the "top secret" seal and gave me 7 pages of "Documentary proved evidence on cases of chemical ammunition scuttling in the Baltic Sea" for initiative research.

Upon consultations with our military experts and scholars from the Russian Academy of Sciences, I wrote a small monograph where I described the ecological threat to the region and suggested possible solutions. My initial research draw attention of the prominent British writer Phillip Knightley, who used it his publication "The Dumps of Death" in *The Sunday Times Magazine* [2].

Documents we have show that the information admiral Tributs shared with me was true. Initially it was agreed that all CW discovered in East Germany would be scuttled in the Atlantic Ocean, 200 miles to the north-east of Faroe Islands, at the depth of 4 km. However, the Soviet merchant fleet was not ready to perform this extremely complicated transport expedition due to the absence of special vessels. Neither the Baltic Navy had cargo vessels for safe long-distance transportation of the chemical weapons.

Under these circumstances, it was decided that the German CW were to be scuttled not in the Atlantic Ocean, but in the Baltic Sea. Soviet troops freighted 6 supply ships with German crews in the British occupation zone to ship chemical ammunition from the Wolgast temporary depot. Additionally, the Baltic Fleet provided two trawlers. About 5,000 tonnes of chemical ammunition and technological containers with CW were delivered to the Eastern part of the Baltic Sea during May 1947 - December 1947 and scuttled there. These CW now lie at a depth of 100-105 m in the area to the north-east of Liepaja at a distance of 70 miles. The total area of the site is about 1,500 sq. m.

Later on, to expedite the operation on CW scuttling, the Soviet troops in Germany chose a new site not far from Bornholm Island. About 30,000 tonnes of CW were scuttled in this area. They lie at the depth of 100-105 m. The total area of the burial is about 900 sq. m. Exact co-ordinates are indicated in our documents.

The remaining 235,000 tonnes of Wehrmaht's chemical ammunition discovered in the American and English occupation zones in Germany had been scuttled together with transports.

At first, all CW were delivered to the ports of Kiel and Emden where temporary storehouses were set up. Then all chemical ammunition was loaded into captured German warships, obsolete British warships, damaged passenger ships collected from all over northern Europe.

About 50 ships of different displacement were gathered. Loaded ships were sailed or towed to the dumping site. Then they were scuttled by explosion or by gunfire. Now we must regretfully acknowledge that the Wehrmaht's CW lie at the sea bottom at dangerous proximity to the coasts of North European countries: at the Norwegian shore of Skagerrak strait near the port of Arendal; at the Swedish shore of this strait near the port of Lysekil; between the Danish island of Fyn and the continent; near Skagen, the northern point of Denmark.

Beside 260,000 tonnes of German chemical weapons scuttled in the European seas by the Allies, the UK also decided to get rid of her own chemical ammunition -- mainly of chemical bombs and shells with yperite and phosgene -- which where scuttled in the Atlantic Ocean.

For over 50 years ships loaded with CW have been lying at the sea bottom. Chemical bombs' shells have rusted by 70-80%, threatening to spread toxic chemicals in the basin of the Baltic Sea.

According to scientific reports, yperite lying on the sea bottom maintains high toxicity for 400 years. Dumped lewisite after hydrolysis will produce the toxic arsenic compounds. Adamsite, chloracetophenone, diphenylchloroarsine and other poisons are very stable to hydrolysis. Nitrous yperite, chloroarsin and some others with hydrolysis will produce secondary highly toxic substances. The poisons will be accumulating in fish and biota. Thus, thousands tonnes of chemical weapons on the Baltic sea bottom are the real danger for the Baltic region.

The famous Russian geneticist Professor V.A.Tarasov conducted a study on the problem of possible negative effect of dumped CW on the European population's health, as most CW are not only toxic but also have a strong mutagenic effect. Stable poisons compounds in minor amounts cause more diverse genetic effects than radioactive radiation. Genetic consequences from CW poisoning are irreversible, and even very small amounts may lead to unpredictable changes in the future generations.

The Baltic Sea's protection started in 1992 in Helsinki when an international convention on this subject was signed. More than 150 pollution sources are listed in this convention's protocols. But even the fullest realisation of the convention will not prevent the ecological disaster if the dumped weapons are not safely isolated.

The Baltic sea is renewing its waters once in 27 years. The biggest threat for the basin is posed by 29,000 tonnes of yperite. More than 1,600 tonnes of yperite have probably got into the sea water already, in a form jelly-like mass covered by a layer of hydrolysis products. In the nearest years about 48,000 units of ammunition (corroded bombs and shells) will start to leak. An effective isolation of CW can be achieved only if it is conducted simultaneously in all six regions of dumping.

On April 7, 1989, about 180 miles to south-west from the island Medvezhiy in the Norwegian Sea, a nuclear submarine *Komsomolets* sank. The studies and experimental works on *Komsomolets*' problem showed that the radioactive sources may be covered by a special sorbent which will absorb them. The sorbent, called *khitozan,* is a gel produced of crabs' tests and can be applied for the CW isolation as well.

The Russian scientists from the of the Karpov Institute of Physical Chemistry (Obninsk Branch) proved that a special composite material may safely isolate poisons. This material consists of a small amount of non-organic gel with a filler like sand or lime. The ash from thermal power stations may be used as a filler as well.

All the dumping areas should be marked and equipped with systems for monitoring and early warning. As the London Convention signed in 1972 does not specify the rules of navigation in the areas of dumping, it seems very important to make a special appendix to it, prepared by international experts.

References

1. Grochler, O. (1978) *Derlaut lose Tod*, Ferlag der Nation, Berlin.
2. *The Sunday Times Magazine*, May 5, 1992.

Section 2.2

Technological Aspects

UNEXPLODED ORDNANCE DEVICES: DETECTION, RECOVERY AND DISPOSAL

DR. RAINALD HÄBER
Heinrich Hirdes GmbH, Berlin, Germany
JÖRG HEDTMANN
SubSea Offshore Ltd., Aberdeen, United Kingdom

1. Introduction

Fifty years after the end of WW II military remains are still a threat to our lives. Dramatic incidents are in the press with astonishing regularity.

The latest spectacular accident happened in Berlin, when a US 500-lbs "Demolition" bomb exploded during building work. It costed the lives of three building workers. This accident emphasised the necessity for thorough site investigations for unexploded ordnance (UXO) and other dangerous left-overs, even for the price of higher development costs.

In principle the same is true for the investigation into sea-dumped chemical weapons, as has been expressed in various national and international reports on the Baltic Sea problem.

The following presentation will highlight some methods of sensor and recovery techniques applied in Explosive Ordnance Disposal (EOD). On the basis of a project application to the European Union we will explain some possibilities for the investigation into chemical munitions on the seabed and finally we will present an innovative and environmentally safe process for the destruction of chemical warfare agents.

2. Sensor and Recovery Techniques for UXO on Land

One has to distinguish between a number of different approaches to the recovery of land-based UXO, depending on:

1. how the munitions had been deployed
    - air raids on targets on the ground
    - artillery
    - infantry
    - static warfare
    - exercises

A. V. Kaffka (ed.), Sea-Dumped Chemical Weapons: Aspects, Problems and Solutions, 73–86.

  - lost or hidden munitions
  - etc.
2. technical solutions for detection and recovery of UXO
    - detection and verification of UXO in specialised recovery team
    - computer-aided acquisition and evaluation of field measurements and subsequent deployment of recovery team
    - multiple co-ordinated bore hole measurements during search for aerial bombs
3. the equipment for detection
    - magnetometer probes
    - induction loops (active, passive)
    - ground radar
    - etc.

An invaluable aid for the detection of aerial bombs and aerial mines is the assessment of aerial photographs, which had been taken during air raids for evaluation. Depending on the quality of the software for the processing of the aerial photography together with careful historical evaluation a relatively high probability can be achieved.

The various sensor techniques for the detection of UXO need to be selected according to the specific problem. Artillery and aerial munitions produced before 1945 were usually made from ferro-magnetic materials. Here a magnetometer probe gives reliable results. Munitions produced after 1945 are increasingly made from para-magnetic or anti-magnetic materials. In these cases the use of probes based on the induction principle becomes necessary.

Particularly on former battle grounds the use of specialised teams seems the most effective solution, since detection and recovery can be combined.

Computer-aided detection and assessment becomes a sensible solution, where surface disturbances have been removed or are non-existent, and deeply buried anomalies are assumed. This is usually the case in excavation pits and in open areas.

The search for unexploded bombs by evaluation of aerial photography must take into account factors like:

- soil conditions at impact site
- height of drop
- speed of attacking plane
- direction of attacking plane
- assumed mass of bomb etc.

In a next step a grid of holes is drilled on the site. During drilling a probe is inserted in the bore hole, to make sure that drill bit does not hit the munitions. The assessment and evaluation of these bore hole measurements usually depends on colour-coded anomaly maps.

3. Sensor and Recovery Techniques for UXO under Water

The detection of munitions on the seabed or buried in the sediment is performed with detection equipment of similar principles as the land-based methods, but, of course, in a specialised waterproof version. In addition, there are a number of technologies available which are specific to underwater work. Some of these will be highlighted in the second part of the presentation.

Because of the size of plan areas, especially near the coasts, often more than one probe are fastened to a sledge or a tow-fish for effective field measurements. Depending on the detection principle the results are often plotted on anomaly maps.

Usually it makes sense to separate the data acquisition from the recovery work, because of the relatively low density of pollution on the sea bed and the different set-ups for the respective operation

Good results can be expected, if more than one detection method are used simultaneously, as shall be described at a later point. These different data sets can be combined to provide unique electronic signatures for the detected targets.

If the sea bed pollution is relatively dense, e.g. under bridges or in harbour areas, it might be more effective to combine search and recovery by using underwater magnets. These magnets should have water jets attached for jetting itself into the mud or sediment on the sea bed. The operating pontoon of the cable dredger, which carries the magnet, should be arranged in such a way that the threat to personnel in case of explosion or leakage is minimised. In 12 years of operating an underwater magnet for the recovery of munitions no accidents have happened.

Of course, a diver could be used for the recovery of the detected munitions where it is appropriate. This diver would usually be assisted through a handheld or helmet-mounted video-system and communication by the surface supervising personnel. However, the availability of cheap and reliable ROV-systems with the appropriate handling systems usually provides the better option, especially in the interest of safety.

4. EMISsO - A Practical Approach

4.1. THE PROPOSING PARTIES

Sub Sea Offshore Ltd.: Underwater engineering contractor - SubSea is the world leader in the design, manufacture, operation and maintenance of Remotely Operated Vehicles and Remote Technology, with an in-house capability in the underwater construction, pipe-laying, installation, hyperbaric welding, inspection and survey, maintenance and repair market. With more than 20 years' experience in the offshore industry and operations in most of the offshore centres of the world, SubSea have expanded their capability by transferring their engineering and survey technology into the environmental field.

Heinrich Hirdes GmbH: Founded 80 years ago, HH historically have their core business in hydraulic engineering and construction, such as piling, dredging, harbour building etc. Originally designed as an internal service to the construction departments, HH have expanded their field of activity into Explosive Ordnance Detection, Recovery and Disposal and the environmental market. For the state of Lower Saxony they have conducted the research into the North Sea ammunitions dumps. Through permanent dealing with unexploded munitions of all kind HH have a comprehensive knowledge regarding all aspects of ammunition.

4.2. INTRODUCTION

As we all know today, the amount of information and reliable data concerning the Baltic Sea dumping of ammunition is rather sparse and of low quality. In a number of meetings during 1993 representatives of the German company Heinrich Hirdes GmbH and Scottish-based Sub Sea Offshore Ltd. discussed possibilities how to reduce this deficit by combining their respective expertise in EOD-matters and underwater engineering. In February 1994, as a result of these consultations an application was filed with the European Union's "Life" environmental program for the "Implementation of an Environmental Monitoring and Information System for Ocean-Dumped Hazardous and Toxic Waste Materials", or in short "EMISsO". The project title and summary do not mention chemical munitions at all. This is intentional as we firmly believe and can prove that our project is applicable to a far greater range of environmental problems world wide. It is this approach that makes an action in the Baltic feasible, since cost can be kept small if there is a wider application for the project results. However, as the dumped chemical munitions are of great concern to all states surrounding the Baltic, we felt that a good example of applying our project would be the Chemical Weapon Dumps, particularly those directly to the East of Bornholm.

The following is a short overview over the aims and foreseen results of the project and the methodology applied. Also, I will emphasise the technological innovations which are to be used during the project.

4.3. AIM OF THE PROJECT

The presently available information about offshore dumps of chemical munitions (CM) and chemical warfare compounds (CWC) do not allow the accurate assessment of their quantities, their location and distribution on the sea bottom, the state of corrosion of the containers and shells, their effect on the benthic environment and data like drift rate, dispersion and sedimentation. Therefore a risk and impact assessment of those dumps on flora and fauna and on the environment as a whole has not been possible with the necessary accuracy. Four national reports from Sweden, Denmark, Latvia and Germany have emphasised the necessity of scientific monitoring of those waste and munitions dumps [2, 3, 4, 8, 16].

"EMISsO" aims **to implement an instrument for the required monitoring and subsequent distribution of information** and at the same time to **actively reduce the deficit in available data** material about munitions dumps.

By combining the technical possibilities of sub-marine data acquisition from other industries and the military field it will be possible to thoroughly investigate the above mentioned parameters in a relatively short, thus cost-effective way. Unmanned equipment carriers will be deployed from a specialised offshore support vessel to gather electro-magnetical, acoustical, visual, chemical and biological anomalies in quantity and quality within a limited survey area.

As a first step on board the vessel the gathered data will be pre-processed, then transferred into a geographical information system (GIS) From this GIS a thorough analysis and subsequent evaluation will take place, from which conclusions can be drawn and recommendations regarding further actions in the dumping areas can be given.

4.4. SCOPE OF THE PROJECT

Within the scope of the proposed project it is envisaged to:

- Establish the dumping site monitoring characteristics for a small section of the Bornholm Basin east of the Danish island of Bornholm in international waters
- Design suitable arrays of seabed monitoring sensors
- Adapt the equipment carriers to suit the above sensor arrays and the required survey parameters
- Design the necessary data structures and software requirements for the implementation of the geographical database and the automatic target detection and characterisation
- Manufacture and test all of the above equipment
- Mobilise the equipment to the sea area from the port of Rostock and identify the site to be surveyed
- Take a series of sweeps with the towed sensor arrays and gather all physical, chemical and biological data
- Process the data to give a mathematical representation of the dump sites and its individual targets
- Based on the model, establish the automatic target detection and characterisation verified by a number of close inspection by ROV
- Return to Bornholm Basin after one year and take a second series of sweeps with sensor arrays optimised according to results from the first survey
- Process the second data set, determining the time lapse changes
- Evaluate and disseminate the results.

The feasibility of such time lapse surveys has been demonstrated in the field of marine research and particularly in the oil- and gas-industry. It is therefore seen as the most suitable, indeed only way to gather enough information about the Baltic munition dumps.

4.5. OFFSHORE LOCATION

For the project we have chosen to concentrate our survey activities on the sea-area called Bornholm Basin, as this is the biggest known dump site.

Depending on the site-identification survey in the area, the chosen offshore location is expected to be a 2 km by 2 km section within the above sea area lying approximately 55 km East of the Danish island of Bornholm in depths between 70 and 130 m.

4.6. EQUIPMENT TO BE USED

The field measurements will be carried out from a specialised offshore support vessel with moonpool. The ship will be equipped with the following apparatus:

4.6.1. *Navigation / Positioning*

- *Differential Global Positioning System (DGPS-) Navigation system with reference stations suitably located*: The Differential Global Positioning System uses up to 10 of 21 available satellites specially deployed in space for navigation purposes. The artificially introduced error for civil applications, which can amount up to a few hundred metres, is corrected by transmitting over an HF-link the differential correction signals which are calculated from one or more reference stations with a known position. The DGPS equipment on board the vessel can thus use those correction factors and together with the information received from the satellites the calculation of the exact position is possible.

- *Acoustic transponder array for the fixation of relative co-ordinates of the survey area:* For underwater navigation an acoustic transponder array will be deployed which allows the tracking of the subsea equipment carriers by long base line calculations. In addition a short base line system exists with transponders mounted fore and aft underneath the vessel so that position relative to vessel can be determined as additional reference.

- *Wide Scan Echosounder:*. The Wide Scan Echosounder allows the scanning of the area in front of the vessel and the equipment carrier in order to circumnavigate obstructions like wrecks or similar. This is necessary due to the low height of the sensor carrier above seabed (approximately 2 m) and the aimed for high survey speed (approximately 2.5 kn).

- *Vessel dynamic positioning (DP-) system interfaced with the above systems:* The vessels DP systems integrates the various navigation systems into an automatic positioning system that controls the vessels thrusters to hold a particular position within very small tolerances or move it along a selected line. These systems are state of the art in the oil service industry and have become so reliable that they ace frequently used even in close proximity of offshore installations instead of time-consuming anchors.

The above navigation and positioning systems allow positioning within approximately 1 m and a repeatability of 1 m. This accuracy is prerequisite for the usefulness of the geographical information database system with individual targets.

4.6.2. *Detection of Anomalies*

The equipment to be used for the detection of anomalies will be a combination of active and passive sensors. Some of these can be described as standard for UXO-detection, others have their place in pipeline inspection or other fields of the oil- and gas-industry. The combination of these different modules is a major part of the uniqueness of this project:

1. *TSS-340 Active Induction Coil System:* Large coils radiate magnetic fields which induce an electric current into electric conductors, e.g. all metallic objects. These currents in turn produce a secondary magnetic field which itself induces a current into the original coils. These currents are detected and produce a typical signature for individual target objects. The size of the current is an indication for the size of the object.

2. *Multi-Magnetometer Array:* This system measures anomalies of the Earth's magnetic field which are caused by ferro-magnetic objects. The display of those anomalies is a direct indication for size and buried depth of the detected objects. The parallel operation of a number of identical magnetometers which central position is tracked by acoustic transponders produces survey data which can be evaluated instantly. This on-line system allows for an exhaustive acquisition of magnetic anomalies of the seabed. The depth of ferro-magnetic objects in the sediment can be measured simultaneously.

3. *Side-Scan-Sonar:* The Side-Scan-Sonar is particularly suited for depths up to 300 m. It operates on two acoustic frequencies so that a high resolution (better than 5 cm) and a large area survey (up to 800 m simultaneously) can be achieved. The equipment is compensated for depth variations so that the acquired data can automatically be corrected. It operates as an acoustic remote sensing system allowing images of the seabed to be displayed and recorded on magnetic tape or as computer data. The principle of operation is to send a sound pulse from the tow fish and record the time and intensity of response for any predetermined offset distance. These are recorded and displayed as a series of grey scales. Port and starboard images can be gathered simultaneously. Side scan sonar is usually used as a seabed mapping tool.

4. *SonarGraphics 3D Mapping System:* SonarGraphics is a high precision 3D mapping system using a controlled high frequency narrow beam sonar. Each sonar "ping" is provided with a full compensation package comprising heading, pitch, roil and x,y,z corrections. The data is displayed on a screen as a contour-plan, profiles or 3D image, all of which are under direct operator's control. A volumetric calculation of detected objects is possible within 5 % accuracy and a resolution of 5-10 cm in real-time. In this high density mode this system produces approximately 80 returns per square metre. The system will be used in linear mode in which it will scan along an axis perpendicular to the towing direction. The direction of tow will provide for the forward axis. This system has been developed in conjunction with British Gas and has successfully been tested in various environmental surveys.

5. *Sub-Bottom-Profiler:* This system is another multi-frequency acoustic system which is comparable to a seismic monitoring system of low seabed penetration capability. The acoustic emissions are frequency-modulated and the pulse rate is adjustable to suit the local requirements. The echoed signal can be recorded and processed into a display-form similar to the side-scan-sonar. It produces information about the thickness of sedimentation and can plot the contour of the seabed below the sedimentation layer.

6. *Chirp-Acoustic-Profiler*: This system is a development of a sub-bottom-profiler, and can - due to the use of external hydrophone arrays - achieve resolutions of 8 - 10 cm. The data is being displayed in real time and gives a multi-colour representation of the layering of the sediments on the seabed. Objects buried into the sedimentation layer will resolve with the same accuracy so that size and shape can be evaluated. The datasets are available in a variety of formats so that compatibility with other data sources can be achieved

7. *Three-Dimensional Camera System:* A visual system that utilises two high-resolution CCD-video cameras. The geometry of the system is known precisely. The two pictures are displayed with through polarisation filters onto one screen. By wearing special glasses the operator is able to reproduce the 3D-impression thus being able to assess objects in a much more reliable way than with simple underwater-video. At the same time the picture information is digitised and a special software allows for dimensional control of detected objects within a few millimetres' accuracy. This system will primarily be used on board the ROV to verify findings with the other sensors.

8. *ART-Laser-Scan-System:* This is a new development, utilising a laser line scanner towed along on the equipment carrier. These laser imaging systems are the most promising technology for underwater long-range vision systems. A synchronous scanner illuminates the target with a narrow laser beam. The target reflections are viewed with a very narrow field of view at the detector, thus only a

very small area of the target is illuminated at any given time. The usual backscatter effects from floodlighting are almost completely eliminated and thus the visual penetration of the water greatly enhanced. The reflection data is digitally processed and image processing techniques can be applied to further enhance picture quality. The result is a near photographic image of very high resolution and defined geometry. It therefore allows for measurements of targets.

9. *Seabed-Gamma-Spectrometer:* The Seabed-Gamma-Spectrometer allows for the monitoring and characterisation of background radiation levels. Such radiation levels are disturbed by substances alien to the surveyed environment and by calibrating the SGS with samples of the wanted substance one can detect pollution from leakage or deliberate discharge.

10. *Laser Fluorimeter:* As a source of biological information we propose the use of a multi-station (up to 12 sampling locations) towed sea water laser fluorimeter for water quality analysis specific to selected hydrocarbons which might be present in the area. The laser excites elements of the plankton population and that of calibrated hydrocarbons (e.g. breakdown products of munitions contents) present in the water. The fluorescent spectra are received through a fibre optic cable, split and counted through specific filters. From this data a direct correlation of the effects of pollution on the plankton population can be made. The system would be towed in conjunction with the multi-sensor towed array.

In addition seabed and water samples will be taken and analysed for correlation with the data gathered by towed sensors.

This multitude of different sensors is necessary for the evaluation of suitability for such survey campaigns. Only after careful assessment of the gathered data a decision can be made which of the described sensor systems and in which combination provide the **most effective identification and characterisation system for waste detection** and monitoring **without compromising accuracy and reliability**. This will then lead to an improved set-up being used for the second survey campaign during the later phase of the project.

4.6.3. *Support Equipment*

1. *Water Jetting System*: Mounted on the ROV, the water jet will allow for the de-burial of detected objects in order to verify the remotely sensed information by close visual inspection and dimensional control with the 3D-camera system.
2. *Drop-Corer*: Will be used to take seabed-sediment-samples from selected areas.
3. *Benthic Grab*: Also used for seabed sampling, but with less penetration as drop corer.

4.6.4. *Equipment Carriers*

1. *Tow-Fish*: For the first time in an environmental survey the SubSea FOCUS ROTV (remotely operated towed vehicle) will be used. This is a specially

designed and manufactured equipment suited to carry all of the planned sensing apparatus. It is laid out in such a way that interference between the various sensor systems is minimal. The system is fully steerable and will be controlled by a pilot on board the support vessel and assisted by automatic height-above-ground steering facility in order to provide repeatable data gathering conditions.

2. *Remotely Operated Vehicle (ROV):* A PIONEER work ROV will be used to carry water jetting equipment, 3D-cameras and sampling tools. The ROV can be directed to the required location very accurately and fast, ensuring the effective conduction of the verification and sampling process.

4.7. THE INNOVATIVE FEATURES OF THIS PROJECT

This project is innovative as it intends to implement a technology that could enable the marine research community and legislative bodies to base their decisions and recommendations on the existence of a detailed database, which contains all necessary parameters for a status assessment of individual waste targets and can be updated in regular intervals to cater for dynamic data. In addition it contains a number of innovative technical and methodological features which ace described in detail as follows:

- The complete data acquisition campaign will take place by the use of unmanned equipment carriers, such as ROTVs (Remotely Operated Towed Vehicles) and ROVs (Remotely Operated Vehicles). These will be deployed from a specially equipped support vessel which can position itself accurately over the intended survey area.

- The planned project features a multi-sensor approach as it will gather data on physical, chemical and biological anomalies over a wide range of frequencies and parameters in a single measuring campaign.

- From the multitude of then available data a characteristic digital signature will be calculated which will be stored into a geographical information system together with position information. The availability of these signatures allows for subsequent monitoring campaigns to be carried out with automatic target detection and identification. The conduction of future monitoring projects thus can be made very fast and cost-effective.

- By repeating those surveys at certain time intervals, time lapse data will be provided. From this one can carry out mapping of static and dynamic parameters such as rate of corrosion, drift, sedimentation, leakage, effects on the marine fauna and flora and dispersion of pollutant. Armed with such accurate information, which is not available to the relevant marine scientists and public bodies today, better judgement of the situation, adequate legislation and possible technical solutions for remedial action will ensue.

4.8. TRANSFER OF RESULTS

Although the project aims to focus on the problem of toxic munitions dumps in the Baltic Sea, the proposed monitoring system and the resultant database are directly applicable to dumps of industrial hazardous and toxic wastes and even nuclear material dumps. The only limitation of the proposed system will be that it will not be suitable to monitor liquid discharges of high solubility due to the high dispersion rate of fluids in water.

However, it is estimated that a more serious environmental threat originates from dumps of solid wastes or wastes packed in containers. The reason is that these containers will eventually corrode and then discharge a relatively large quantity of very concentrated material into a very localised area, thus causing a great danger to the surrounding environment. It is therefore of particular interest to find, characterise and assess those dumps before their state becomes a real threat or even a catastrophe.

5. The Recovery of Chemical Munitions from the Seabed

5.1. INTRODUCTION

When talking about the problems which might arise from possible recovery operations one often neglects the fact that there is a huge potential of technologies available from other industries, especially those which have invested massively over the past two decades. Of particular interest are the offshore oil and gas industry, the nuclear industry and the military.

We will demonstrate how some of these resources can be combined to a logical concept for the recovery and disposal of chemical munitions dumped in the Baltic Sea.

5.2. METHODOLOGY

A project comprises three principal parts or phases:

1. site investigation
2. recovery
3. disposal.

5.2.1. *Site Investigation*

This phase has been covered in the previous section. It is assumed that the results and the experience from "EMISsO" will lead to all necessary procedures to conduct phase 1 in a fast and cost-effective manner. It is the evaluation of data from this phase that will determine whether to initiate phases 2 and 3 at all.

5.2.2. *Recovery*

If it is decided after phase 1 to go ahead with a recovery program, targets for recovery will have to be determined. For that purpose targets will have to be classified according to their physical condition, i.e. state of corrosion, depth of burial, physical size, type of shell etc. The technologies for a recovery will have to be adapted to suit targets within their respective classes. A prime concern, however, will be the recovery of fundamentally intact munitions, i.e. not free lumps of agent.

If munitions are found within ships hulls or other structures, a different approach with modified technologies has to be adopted. HH and SSOL have available concepts for such a situation, however, for the purpose of this lecture we will concentrate on "loose" munitions.

The identified targets will be approached by ROV's which are equipped with specialised tooling for the safe handling of munitions. These ROV's will place the munitions within a disassembly-container, which can be sealed from the surrounding environment and is also pressure-proof. Inside the container a dedicated tooling-package places the munition onto a table, where it is pre-processed. The method of processing depends on the type and the state of the found munition. Shells could be cut using abrasive water cutting, if the case was still very much intact. Alternatively the whole shell could be placed into an anolyte bath for complete destruction. This option would be obviously desirable, if munitions seem to be in an unstable condition.

If the shells are cut open by whatever method, the contents has to be flushed out. This technology is in daily use in the recycling of conventional munitions and has proofed reliable. The resulting suspension of chemical agent, explosive, abrasive and water as well as the anolyte bath can be fed into a transportation container which will carry the mixture to the surface support vessel, where it is docked onto the processing container. There the necessary decompression will take place in a controlled environment. This step is very similar to the transfer of deep-sea divers from their work place back to their pressurised accommodation chambers by means of a diving bell. The system is proven numerous times in the offshore oil industry and is very safe. Also, the danger of inadvertent leakage of agent into the environment because of pressurisation whilst on the seabed is minimised, as opposed to direct recovery to the surface.

In shallow water, however, the effects of pressurisation are less likely to be dramatic, as the experience shows. Here the munitions can be recovered through a sealed moonpool area directly into a surface processing plant.

In any case, the need for intermediate storage with all the connected problems of safety and CW Convention regulations is eliminated by taking the destruction process to the problem.

5.2.3. *Disposal*

On board the support vessel an electro-chemical oxidation system will be installed which will destroy the organic compounds recovered from the seabed. This system is tightly sealed from the environment and can be operated with slightly negative pressure as an additional safety feature. The transfer container can be docked onto the e.c.o.s. plant and the chemical warfare agent together with the scrap metal is then transferred into the process. Within the process plant the organic compounds are completely and safely destroyed and the scrap metal is scoured to remove any CW residues.

The Dounreay Electro-Chemical Process. Core of the destruction of chemical warfare agents is a transportable electro-chemical oxidation plant, better known as the Dounreay-Silver-II Process. This process has been developed by the UK Atomic Energy Authority (AEA) and has successfully been tested on a wide range of organic compounds, including explosives and propellants and chemical warfare agents like mustard, viscous mustard, Sarin and VX. Also the combined destruction of VX and explosives together with the complete shell has successfully been demonstrated in the UK CBDE in Porton Down, England. A pilot plant is situated within AEA's research centre in Dounreay, North Scotland.

The process has a number of advantages over destruction processes, particularly incineration:

- It is possible to feed explosives and chemical agents simultaneously. These are then, as all other organics, oxidised to carbon dioxide, water, nitrous oxides and mineral salts. The system has been praised in several publications for its built-in safety and environmental compatibility.
- The destruction takes place at low temperatures, typically between 50 and 90 degrees, it is therefore impossible that toxic by-products commonly associated with incineration (e.g. dioxins, di-benzo-furanes) are formed.
- The process operates at ambient or slightly reduced pressure. This minimises the risk of accidental emissions.
- The system can be switched off and rendered inert within seconds by simply isolating the electrical supply.

The breakdown products are inorganic and relatively harmless. They can, if required, be further treated by standard processes.

References

1. *Mehr Giftgas in der Ostsee als angenommen,* Berliner Morgenpost and other daily newspapers from 14.02.94 (quoted from an interview NDR-TV with Prof. Ruhl from BSH on 13.02.94).
2. *Bundesamt fur Seeschiffahrt und Hydrographie: Chemische Kampfstoffmunition in der sudlichen und westlichen Ostsee, Bericht der Bund/Lander Kommission "Chemische Kampfstoffe in der Ostsee",* BSH Hamburg 1993.

3. *Sjofartsverket: Rapport om kartlaggning: av forekomsten av dumpade kemiska stridsmedel pa den svenska delen av kontinentalsockeln,* Norrkoping 1992.
4. Laurin, Fredrik (1993) *The Baltic and North Sea Dumping of CW - still a threat?* SIPRI.
5. *Ergebnisbericht: Untersuchung von Munitionsversenkungsgebieten in den niedersachsischen Kustengewassern,* Heinrich Hirdes GmbH and BBS Brandenburg, Berlin 1992.
6. Rapsch, H.-J. (1992) *Ist die Nordsee eine Rustungsaltlast?,* Abfallwirtschaftsjournal **4** NI. 2.
7. Evstafiev, I. B. et al. (1992) *The Mechanism of Corrosion of the CW-Shells Dumped in the Sea-Water,* Kiel.
8. Danish National Report on the Dumping of Chemical Ammunition in the Baltic Sea, 1992.
9. Latvian Report to the Helsinki Commission's Working Group on Dumped Chemical Munitions (HELCOM CHEMU).
10. Knightley, Philipp, *Dumps of Death, Sunday Times*, May 1992.
11. *Die Reportage - Ostsee - Giftsee?* TV-Report from Peter Berg, 10.07.92, ZDF-Television, Germany.
12. Krohn, Axel (1993) *Eine neue Sicherheitspolitik fur den Ostseeraum*, Opladen.
13. Fonnum, Frode (1993) *Examination of CW amminition in the Norwegian Sea - Possible Environmental Impact*, Norwegian Defense Research Establishment.
14. Narewski, Marek, University of Gdansk (1993) *Visual Inspection of War Gas Deposits by ROV's Methods,* Kiel.
15. Several Datasheets and Information Leaflets from Equipment Manufacturers.

THE TECHNOLOGICAL PROBLEMS WITH SEA-DUMPED CHEMICAL WEAPONS FROM THE STANDPOINT OF DEFENCE CONVERSION INDUSTRIES

DR. V.N. KONKOV
Chief Designer, Contech Scientific-Industrial Corporation
Kaliningrad, Moscow Region, Russia

The Russian military-industrial complex (VPK, in Russian abbreviation) is currently looking for every opportunity to use its scientific and technical potential and the experience gained in the realisation of complex comprehensive programs. But economic conversion proceeds slowly, mainly because it does not take proper account of specific peculiarities in the Russian VPK which are unique not only due to the size, but also due to the super-specialisation of the majority of its enterprises, that are most often research, experimental design or product test structures. The parity reached in a number of modern weapons was conditioned not only by the production potential of the country, but also by the proposed original approach in tackling research and design problems. A unique school of engineering thought has formed in Russia, able to deal with complicated problems, including those of global ecology.

When the problem of captured chemical weapons sea-dumped by the Allies after the end of World War II became actively discussed in recent years, it turned out that the Russian VPK possessed certain possibilities to attack the key tasks of the problem: reliable diagnosis of the state of the environment and dangerous objects and, when necessary, their safe lifting and destruction.

The proposals were based on successful experience accumulated over many years of cryogenic and thermochemical missile technology and shipbuilding for the Navy.

While official data is not available, different Russian sources - which were also using information from foreign media - report that from 1946 to 1948 in Skagerrak, near the Norway and Sweden shores, the British and American military sea-dumped 50 ships with 270 thousand tonnes of captured chemical weapons. About 30 thousand tonnes were dumped within an area of 2,000 sq. km near the Gottland and Bornholm islands by the Soviet Army.

Publications have appeared on chemical weapons sea dumps in other regions of the world's oceans: the White and Barents Seas, in the Indian and Pacific Oceans, including regions near the shores of the USA and Australia, in the Biscay gulf of the Atlantic and in the Irish Sea.

Barrels, cisterns, aircraft bombs and shells filled with 15 types of toxic agents, including the organic-chlorine mustard gas, organic-phosphor tabun and sarin, arsenic

A. V. Kaffka (ed.), Sea-Dumped Chemical Weapons: Aspects, Problems and Solutions, 87–91.

lewisite and adamsite have been enduring the sea water action for almost 50 years. The accidental casting of those objects onto the Scandinavian sea coast or their fishing out by Baltic fishermen provided the possibility to establish the degree of the shells destruction as critical.

The data on possible CW migration rates in sea water, hydrochemical processes of their interaction and toxic danger of the products of such reactions is controversial. Suggestions concerning the feasibility of lifting and neutralising the dumped chemical weapons are also poles apart. The positions of those participating in discussions are determined by not only and not so much ecological and techno-economical considerations, but by political and salaried ones. The opinions of the countries directly interested in this problem, especially the Baltic states and the former Allies, are ambiguous as well. The far-too-high estimated cost of the suggested lifting operations and the extreme requirements for their safety determine the reserve in the position taken by officials, though Russia was the first to announce the problem and her readiness to participate in its solution.

At the same time, the scientific community and movements of environmentalists are expressing apprehension for the advisability of the passive expectation of further developments, considering the fact that the corrosion process in the main types of dumped CW carriers (thin-wall containers, barrels and bombs) is nearing or has neared its end. Specialists in chemistry and toxicology, including genetic toxicology, are sounding an alarm at the fate of the basin, the shores of which are populated by 80 million people.

Two principally different methods were used in dumping chemical weapons in the Baltic Sea:

- concentrated dumping of containers, shells and bombs together with the ships;
- dispersed dumping of individual containers, shells and aircraft bombs from moving vessels; altogether 500,000 objects have been dumped by the Soviet Army.

The dumping depth in both cases is 50 - 100 meters.

Several years ago, Contech scientific and technical corporation proposed new approaches to deal with the concrete tasks of diagnosing the condition and the safe lifting of CW dumped objects. The new methods are based on a cryogenic technique for freezing samples and small objects, and the utilisation of cryogenic pontoons for raising large objects. Specialists in missile technology propose to use the experience of installations for the neutralisation of highly-toxic missile fuel components to create floating complexes for destruction of organic- chlorine and organic-phosphor CW.

The following possible areas of application for our technologies are suggested:

(1) Sample-taking

Monitoring of dumped objects, especially dumps of the second type, involves numerous takings of water and bottom silt samples, and when necessary, the recovery of a limited number of objects for objective assessment of the CW and its envelope's condition (thick-wall shells and thin-wall containers and aircraft bombs). There is information that the weapons may also contain explosives in their composition, so additional investigation is required.

To solve the problem of mass control of the condition of the environment in the CW dumping areas, a principally new method of batometry was proposed, i.e. automatic water and soil sampling by a diving apparatus with the freezing and subsequent laboratory analysis of the sample. The promptness of the method is determined by the fact that many samplers are lowered from a ship or a plane, and then they surface according to a pre-set programme to be collected. This allows permanent control over vast areas, and when necessary, a prompt change in the programme of operations.

Utilisation of an inert liquid gas as a source of cold enables simultaneously to solve the problem of getting the necessary amounts of gas and to control buoyancy, while keeping the pressure in the compartments equal to the hydrostatic one. That allows the production of a simpler and cheaper submersible. The apparatus becomes more like a space rather than deep-water craft and consists of a cryogenic thermoinsulated tank containing liquid nitrogen, a heat exchanger and elements of pneumohydroautomatics, power systems and light, sound and radio beacons.

After submersion, the cryogenic tank becomes warmer, the tank pressure increases to reach the value of hydrostatic pressure. When the required depth is reached, liquid nitrogen is fed to the heat exchanger - freezer. It takes 1 litre of liquid nitrogen to freeze one litre of water (without considering losses). When the ballast is dropped, the apparatus moves upwards, constantly releasing the excessive gas to level the tanks pressure with the pressure outside. The removal of the sample takes place onboard the research ship, after the heat exchanger warms up.

(2) Raising small objects

Further development of this method resulted in cryocontainering and raising such relatively small objects as cisterns and aircraft bombs with the help of an independently emerging apparatus. Additionally, the technique solves the problems of automatic searching for the object from the submerged apparatus, a situation close to that of spacecraft docking.

The scale and relative complexity of such a project - including the refitting of research ships for the CW to get new containers - requires the co-operation of space, rocket and shipbuilding firms, what used to be traditional to them when they were responsible for defence program implementations.

(3) Raising ships loaded with CW

The third direction could include the development of technology for raising sunken ships with dangerous substances aboard.

According to eye-witnesses, the ships in the port received a cargo of chemical weapons, the holds were sealed with concrete, the hatches were welded up, and the ships were towed to their destination, where they were sunk. They were passenger ships, mine sweepers and motor torpedo boats.

The diversity in the ships types and dumping conditions (depths, sea grounds, presence of silt or bottom currents) prevents the use of a single universal facility for ship raising, but the principle should be common and based on the creation of a powerful icy foundation under the object to be raised.

One way of doing this is a submersible pontoon (bell) that is lowered onto the ship from above. When the pontoon is sitting on the sea bottom, drilling devices start simultaneously drilling several dozens of holes in the seabed. Liquid nitrogen from the pontoon-installed cisterns is pumped through the drills, water forms ice on the drills and in the drilling holes and breaks the ice foundation away, even from rocky ground.

Formed by sea water, the icy bottom seals the pontoon, and when the cisterns are empty they act as buoys and take the pontoon away from the sea bottom together with the ship. In day time, the pontoon is towed to a special dock which is the very place where all the operations of CW removal take place (making use of safe methods of dismantling, draining of fluid products, neutralising of assembled steel sections typical of atomic and missile technologies).

(4) Decomposition of CW

The thermochemical decomposition of organic-chlorine and organic-phosphor CW in oxygen environment (these account for the absolute majority of the Baltic dumped chemical weapons) permits them to be reduced to some easy-to-decontaminate or harmless compounds like carbon dioxide, phosphorus oxides or hydrogen chloride, sulphur oxides. The processes take place in combustion chambers at 3,000-3,500°C and a pressure of several MPa, using methods traditional for liquid-propellant rocket engines. The end combustion products of other CW might be additionally condensed to trap the hazardous compounds.

The floating complex for CW neutralisation appears to be the only acceptable option, as neither the former Allies nor the Baltic states will agree to have such production on their territory.

Though they look exotic, all mentioned methods/approaches and operations are based on actual experience whose accumulation has taken a lot of time and means. This allows us to proceed with the experimental designing of manufacturing demonstration models of the required equipment.

The cost estimate of projects mentioned by different authors (from 20 to 80 billion US dollars) is up to now based only on intuition. If a project design was made, it would be hardly probable that the cost of the program exceeded 2-3 times the cost of the American program that took men to the moon. One thing is clear: not a single state involved in this Baltic problem can deal on its own with the entire complex of political, ecological, technological and economic problems.

At the same time the creation of such a complex would be a powerful impetus in the development of a number of imperative directions in modern science and technology. The elements to be created for the complex are also needed in industrial ecology (elimination of hazardous toxic waste), marine oil and gas production, hydrogeological exploration of marine deposits, sea transportation of liquefied gases (methane, nitrogen, oxygen), etc.

The Contech corporation is interested in co-operating in the work under the present program and, most of all, in underwater cryogenic monitoring, giving an investor broad opportunities to use development materials, equipment and patent and licensing rights. The total expenses for the execution of demonstration models do not exceed 200,000 US dollars.

TECHNOLOGICAL QUESTIONS OF SAFE ELIMINATION OF CW DUMPS ON THE BALTIC SEA BED

PROFESSOR L.P. MALYSHEV
Ministry of Emergency Situations of Russian Federation
Moscow, Russia

During the forty-six years of the Cold War the natural purificatory abilities of the world's oceans and land were mercilessly exploited.

The coastal depths of seas in high-technology countries were turned, over the Cold War years, into dumps of highly toxic radioactive waste and grave yards for sunken military and civil ships. Over 680,000 containers with radioactive waste have been dumped by twelve countries, with Russia accounting for under 10% of the containers. There are five sunken atomic submarines resting on the ocean bottom, 7 reactors and up to fifty pieces of nuclear ammunition. Among them there are three nuclear-powered vessels, five reactors and 25 units of nuclear ammunition of Russian origin [3,4].

It has been documented that four countries (Great Britain, the USSR, the USA and France) carried out the burial of chemical weapons belonging to fascist Germany. However, only the burial of 302,000 tonnes of CW in the Baltic Sea, performed by Great Britain and the USSR, has detailed description [5]. Britain dumped another 120,000 tonnes of its own CW in the English Channel [6].The open press has enough convincing evidence about the USA having dumped CW near Australia, and the USSR into the White and Japan Seas [7].

On those dumping grounds in the Baltic Sea which were controlled by Russia, there are over 600,000 chemical ammunitions, containing more than 12,000 tonnes of chemical war gases, of which 81% are mustard gas and arsine oil (Tables 1,2).

The main factors determining the dynamics and development tendencies in the situations with chemical warfare agents in the Baltic Sea include the quantity of the sea dumped ammunition, its resistance to corrosion, effect of the physic-geographic conditions on CW and on their hydrolysed forms in the sea water (Tables 3,4).

Investigations conducted by Russian scientists [36,37] in the two CW dumping grounds under Russian control testify that it is possible to expect three periods of tightness failure of the chemical ammunition. Each period when mustard gas will leak into the sea water will create an extreme situation (Table 3).

The first period is expected to begin 50 years after the dumping, i.e. from 1997, and to last for 26 years, with mustard gas leaking at a rate of 43.3 tonnes per year or 700-800 tonnes of mustard gas during the entire period.

A. V. Kaffka (ed.), Sea-Dumped Chemical Weapons: Aspects, Problems and Solutions, 93–104.

TABLE 1. Amount of warfare munitions dumped in the areas controlled by USSR (Russia)

Types of chemical munitions, units	Munitions with mustard gas	Mun. with As-CW	Mun. with Adamsite	Mun. with CAP	Mun. with other CW	Total, units
Aircraft bombs	5,960	721	639	414	-	7,464
Aviation smoke candles	-	-	2,790	-	-	2,790
Artillery projectiles	26,205	-	2,564	4,300	-	33,069
High-explosive bomb	2,720	-	-	-	-	2,720
Mines	830	-	-	-	-	830
Barrels	42	124	-	-	48	214
Containers	-	80	-	-	-	80
Drums	-	-	599	-	-	599
Jars	-	-	-	-	626	626
Total, in the 1st region	**35,487**	**925**	**6,592**	**4,714**	**674**	**48,392**
Aircraft bombs	65,779	8,338	7,388	4,785	-	86,290
Aviation smoke candles	-	-	32,250	-	-	32,250
Artillery projectiles	302,926	-	26,639	49,702	-	382,267
High-explosive bomb	31,442	-	-	-	-	31,442
Mines	9,590	-	-	-	-	9,590
Capacities (barrels)	487	1,434	693	-	552	3,166
Containers	-	924	-	-	-	924
Drums	-	-	6,927	-	-	6,927
Jars	-	-	-	-	7,234	7,234
Total, in the 2nd region	**410,224**	**10,696**	**76,897**	**54,487**	**7,786**	**560,090**
Aircraft bombs	71,469	9,059	8,027	5,199	-	95,754
Aviation smoke candles	-	-	35,040	-	-	35,040
Artillery projectiles	329,131	-	32,203	54,002	-	415,336
High-explosive bomb	34,162	-	-	-	-	34,162
Mines	10,420	-	-	-	-	10,420
Capacities (barrels)	529	1,558	693	-	600	3,380
Containers	-	1,004	-	-	-	1,004
Drums	-	-	7,529	-	-	7,529
Jars	-	-	-	-	7,860	7,860
Total, in 2 regions	**445,711**	**11,621**	**83,492**	**59,201**	**8,460**	**608,482**

The second period is expected 75-80 years after the dumping and should last for 60 years with the mustard gas coming at a rate of up to 280 tonnes per year, which would make up to 6,500 tonnes of mustard gas during the entire period.

The third period (2110 - 2260) is expected to last 150 years with a low mustard gas emission rate - up to 13 tonnes per year.

TABLE 2. Amount of chemical warfare agents dumped in the areas controlled by USSR (Russia)

Types of chemical munitions, tonnes	Mustard gas, t	As-containing, t	Adamsite, t	CAP, t	Other CW, t	Total, t
Aircraft bombs	512	78	51	41	-	682
Aviation smoke candles	-	-	6	-	-	6
Artillery projectiles	58	-	5	3	-	66
High-explosive bomb	27	-	-	-	-	27
Mines	4	-	-	-	-	4
Capacities (barrels)	7	18	60	-	6	91
Containers	-	80	-	-	-	80
Drums	-	-	2	-	-	2
Total, in the 1st region	**608**	**176**	**124**	**44**	**6**	**958**
Aircraft bombs	5,920	906	591	479	-	7,896
Aviation smoke candles	-	-	65	-	-	65
Artillery projectiles	671	-	61	36	-	768
High-explosive bomb	314	-	-	-	-	314
Mines	42	-	-	-	-	42
Capacities (barrels)	80	203	693	-	74	1,050
Containers	-	924	-	-	-	924
Drums	-	-	18	-	-	18
Total, in the 2nd region	**7,027**	**2,033**	**1,498**	**515**	**74**	**11,077**
Aircraft bombs	6,432	984	642	520	-	8,570
Aviation smoke candles	-	-	71	-	-	71
Artillery projectiles	729	-	66	39	-	834
High-explosive bomb	342	-	-	-	-	342
Mines	46	-	-	-	-	46
Capacities (barrels)	87	221	753	-	80	1,140
Containers	-	1,004	-	-	-	1,004
Drums	-	-	20	-	-	20
Total, in 2 regions	**7,636**	**2,209**	**1,552**	**559**	**80**	**12,036**

The problem of averting emergency situations and providing ecological safety in implementing programs for the safe elimination of chemical war gases on the Baltic Sea bed touches upon the interests and safety of the entire world community [28,39].

It seems that the settlement of the problem to avert emergency situations and to ensure ecological safety during the elimination of CW dumps would be most reasonably organised in two stages and under the supervision of an inter-governmental control body consisting of representatives of the Baltic states, Russia and other developed countries, participants of the Chemical Weapons Convention

The specific features in the creation of safety systems while working with chemical warfare agents include, firstly, the difficulty in searching for the silt-covered munitions

TABLE 3. Predicted dynamics of mustard gas penetration into sea water in the dump areas controlled by USSR (Russia)

Region 1

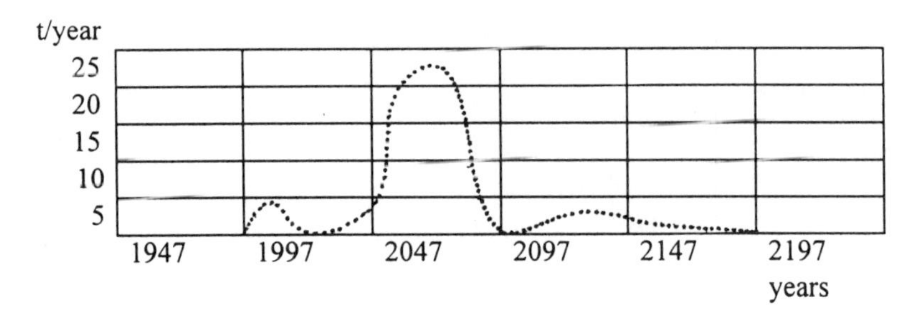

Region 2

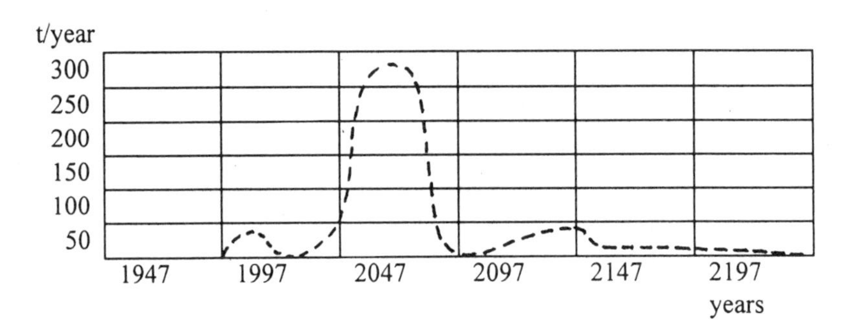

Total, Regions 1 and 2

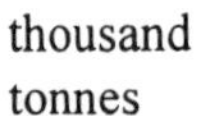

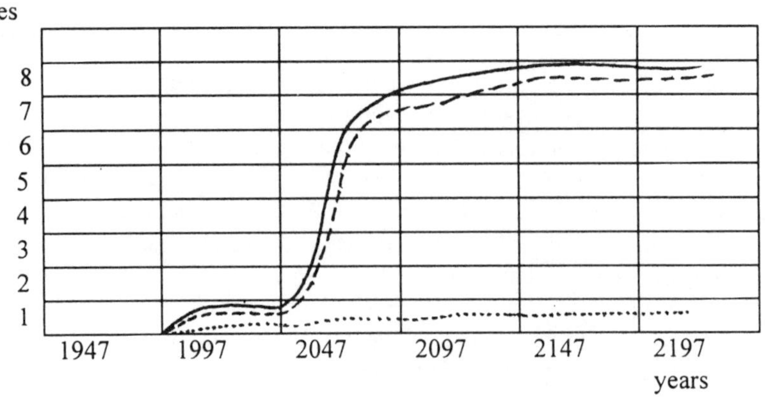

on the sea bottom, although their dumping grounds have been found and outlined by marine expeditions. Secondly, the difficulty in complex instrument indication of the hydrolysed forms of the decomposition of mustard gas and arsine (over 10 components) against the background of streams of pollutants from the technogenic waste disposals in the Baltic Sea.

Thirdly, during the elimination of chemical war materials, the amounts of resulting chemically hazardous wastes are 5 - 9 times as many as the initial material and they should be reclaimed and buried on the territory of a country that would agree to provide its territory for establishing systems of a long-term controlled burial, because mustard gas remains toxic for more than 400 years.

In the first stage, it is reasonable to create a system of expeditions and rescue-in-emergency parties servicing CW dumps that would include:

- a subsystem for complex observation, early warning and protection of dumping grounds of chemical war materials;
- a subsystem under which expeditions conduct observations and investigations of the situation development;
- a subsystem of rescue-in-emergency to react in extreme situations.

In the second stage, it is reasonable to establish a system for the ecologically safe elimination of chemical warfare agents from the Baltic Sea bed, which would include:

- a subsystem for the collecting, lifting and delivery of chemical ammunition;
- a subsystem for the inactivation of ammunitions, the reprocessing of chemical agents and the utilization of highly-toxic waste;
- a subsystem of complex observation over the safe execution of technological processes and the condition of the personnel, indication of leaks and violations of production cycles in the processes of ammunition inactivation and the elimination of chemical war materials, salvaging of waste;
- development of emergency situations and accidents in the work of the systems of lifting and elimination of chemical war materials;
- a subsystem for the transportation and safe burial of highly-toxic waste.

It seems advantageous to explain in more detail the function and structure of the subsystem for complex observation, warning and protection of the CW dumping grounds, as, in the present conditions of the growing intensity of anthropogenic activity, the chances of unpremeditated and premeditated intrusion of people into the areas with the CW dumping grounds are growing.

As the analysis of the investigation experience of the Institute of Oceanology of the Russian Academy of Science has demonstrated, in the dumping grounds of chemically and radioactive hazardous objects, it is advisable that the subsystem for complex observation over the aquatory would consist of three types of submersible robots.

The first type - submersibles for background monitoring, intended for measuring and recording the speed and direction of the currents, water temperature in relation to

TABLE 4. Basic properties of the dumped chemical war materials in the sea water of the Baltic Sea

CW type	Name, formula, designation	Physical properties	Lethal doses	Sea water solubility	Appearance in sea water	Additional information
S K I N I R R I T	Mustard gas $S(CH_2CH_2CL)_2$ has many names	$t_{melt.}$=14.5°C $t_{boil.}$ =217°C $t_{decompos.}$>500°	Inhalation: 1.5 mg-min / liter at t > 10 Oral route: 1 - 2 mg / kg Skin absorption 70 mg / kg Latent period > 4 days	at 20°C - 0.07% at 0°C - 0.03% poorly soluble with increased water tempera-ture the hydrolysis rate and gas decomposition is increasing	(1) jelly-like substance (2) jelly-like drops on the sand or bed silt (3) hydrolyzed forms on the border of the substance wash-out by sea water (4) thin film on the water surface in shallow sea regions	1. Mustard gas has certain surface activity, which reduces water surface tension, therefore the gas spreads on the water surface, forming a film and boosting gas decomposition. 2. Gas easily mixes up with all warfare agents, benzene, halogenated hydrocarbons, fats and oils, which aids its spreading in sea water. 3. Gas dissolves sulfur easily. 4. Gas has a high seeping characteristic through number of substances, materials and skin. 5. Gas speeds up steel corrosion and gas liberation (hydrogen, hydrogen sulfide, ethylene and other), which increases pressure in capacities, barrels, shells, bombs, etc. 6. Gas possesses cumulative properties. 7. Gas has no antidotes yet.
A N T	Arsine oil mixture of phenol dichloroarsine with diphenylchloro-arsine, trichloro-arsine and arsenic chloride	$t_{melt.}$=40°C density: 1.43 g / cm	Inhalation: 15 mg-min / liter Oral route: 15 mg / kg	at 20°C - 0.2 % at 0°C - 0.1 % poorly soluble	crystal toxic substance	1. When in water, gives transitional forms: diphenyl-arsenious acid and dephenyl-arsyne oxide, also toxic. 2. Hydrolyzes with formation of water soluble salts of dephenyl-arsenious acids.

TABLE 4 *(Continued)*

CW type	Name, formula, designation	Physical properties	Lethal doses	Sea water solubility	Appearance in sea water	Additional information
IRRITANT	Adamsite Ph(NH) (AsCl)Ph sternite, causing irritation of upper air passage	$t_{melt.}$=195°C molecular weight: 277.59	Inhalation: 0.02 mg-min / liter at t > 30 min. 30 mg-min / liter at t < 5 0.002 mg / liter	practically insoluble	crystal toxic substance	Hydrolyzes very slowly in alkaline media with formation of dihydrophenarsazine. Its irritating action is close to that of adamsite.
	Chloraceto-phenone, CAP $Ph(CO)CH_2Cl$ lacrimator, causing tear secretion and skin irritation	$t_{melt.}$=59°C molecular weight: 154.6 density: 1.321 g / cm	non-lethal	low soluble	crystal toxic substance	Remains toxic in water, can be distilled by steam without visible irritation.
GENERAL ACTION	Cyclone-B HCN hydrocyanic	$t_{melt.}$=15°C molecular weight: 0.6969	Inhalation: 0.2 mg / liter at t > 10 min. Oral route: 1.0 mg / kg	quite soluble	water solutions of hydrolysis products of hydrocyanic acid are not toxic	Water is hardly polluted by hydrocyanic acid.

depth stratification, precipitation rates of solid and suspended particles, determination of chemical or radioactive substances in a water volume unit, the level of chemical radioactive pollution of the sea bottom soil in the location of the apparatus, registration of threshold values of the chemical or radioactive contamination and making forecasts of the aquatory contamination intensity, including determination of the threshold values for the beginning of the degradation of the natural and biotic environments. The equipment includes the following microprocessor modules:

- a module with a multiprocessing measure-forecast-control complex with a memory sufficient to support a library with procedures and an archive of registered measurements data and to make forecasts of the intensity of the polluting processes, including determination of threshold levels;
- a module of telecommunication facilities and their control;
- a module of power supply and energy converting;
- a module of fault control in the operation of the submersible.

Proposed by the Russian Academy of Sciences' Institute of Oceanology, the scientific-technological concept of ecological monitoring [40] has been based on the idea of using underwater stations, installed on the sea bed near the submarine "*Komsomolets*" that would be replaced each year. In this modification, the power unit, based on accumulator batteries, can ensure the operation of three modules of the submersible as well as recording of measurement data of the current, temperature and water pollution.

To ensure the operation of the proposed underwater station on a complete cycle of the assignments for aquatory background monitoring, it will be necessary to develop a module of complex power supply, which would include an accumulator battery, an energy converter and an onboard engine - a.c. generator. The a.c. generator of choice will be the one designed by O.A. Mozhaev, which operates on the principle of a vortex generator under the pressure of a water jet with a speed exceeding 1 knot (0.5 m/sec) and overall dimensions of height - 0.12m and diameter - 0.25m to produce 2 kW of power and the dimension of 1.75/5.5m to produce 100 kW. In that modification the submersible is able to perform a complex of assignments of background or object monitoring and to transmit the observation data, using the acoustic or radio channels, to research ships, coast stations or the command post at an agreed time or on the request of the arrived ship after a code signal is exchanged.

For the second type of submersibles (robots intended for object monitoring) it is advisable to include in the number of assignments the measurement of the oxygen content in the water above the source of possible emission from the controlled object, measurement and recording of the water temperature and current velocity, measurement and recording of the concentration of the polluting emissions from each emission source of chemical or radioactive material, registration of the emission threshold values, detection of the symptoms and forecasts of the build-up in the process towards emergency situations at the controlled emission source, activation of subsystems for warning signals, control over the mechanisms that compose and supply mixtures used to slow up the oxidising and polluting processes in the controlled emission source.

Apart from the above mentioned power modules, multiprocessor and instrumentation modules, it will be necessary to install modules for the storage and preparation of degassing mixtures, a module to check the accuracy and safety of mixtures introduction and the stock of chemical agents.

Into the third type of submersibles for the subsystem of observation in the chemical ammunition dumping grounds, it is proposed to include a robotic submersible with navigation referencing and a function for guarding the object. It is intended for:

- the detection any visitor;
- identification, using codes, on request of a "friend" vessel;
- examination of the actions of the vessel or visitor in relation to the guarded object and transmission to the command post of the warning signal, if the danger of the visitor's actions has been established;
- active, demonstrative follow-up on the visitor's actions with recording of an audio or video portrait and transmission of the warning signal.

Many of the proposed modules for submersibles can be developed through miniaturisation of the equipment used in the Russian Navy or based on the developments of R&D companies *"Gidropribor"* or *"Akva-tekhnika"*, which appeared in recent years [41].

The market appearance of the above submersible robots makes it possible to discuss the possibility of creating national or international systems for the ecological monitoring of coastal sea regions.

Thus, preliminary investigations demonstrate the advisability and necessity of developing autonomous and remote controlled submersible robots for the purposes of aquatory (background) and object monitoring, safeguarding of the objects of impending danger, for purposes of mobile observation and control over the safety of ship-raising and rescue-in-emergency operations.

It would also be advisable to organise a competitive selection of technologies for the guaranteed safe dismantling and utilisation of nuclear, radio and chemically hazardous weapon systems.

The preliminary estimate of the radioactive or chemically hazardous sea-dumped objects suggests that chemical warfare agents of the type of mustard gas will remain hazardous for more than 400 years and the ecological impact of radionuclides of plutonium from weapons will constitute danger for more than 240,000 years.

It is imperative to work out a complex state program to ensure the safety of disarmament and conversion activities in Russia, to enhance the control efficiency within the framework of the Russian system of prevention and actions in emergency situations, control over the insurance of all kinds of security, environmental protection and the reduction of possible damage to the country's economy due to the beginning of disarmament processes on military objects.

Literature and References

1. *Information Material on the Problems of Nuclear Disarmament* (1992) Prepared by the organization "Physicians of the World for Prevention of Nuclear War". Moscow-New York.
2. Zhdanov N., (1993) Strategy of Survival is Needed by the Whole World, *Green Cross* **1.**
3. *Facts and Problems, Related to the Dumping of Radioactive Waste in the Seas Washing the territory of the Russian Federation* (1993). Administration of the President of Russia, Moscow.
4. Bukan'S.P (1992) *In the Tracks of Underwater Catastrophes*. Moscow.
5. *Chemical Munition Dumped within the Area of the Helsinki Convention.* Report to the Third Meeting of the Working Group on Dumped Munition. Copenhagen, Denmark. December 1993.
6. Knightley P. (1992) Dumps of the Death, *Sunday Times Magazine*, 5 April, p.28
7. *Glasnost* **21**, 10 June, p.6
8. *USA Militarism. Reference book.* Moscow, IPL, 1985, p.322.
9. Yakovlev N.N. (1988) *War and Peace American Way*, Pedagogika, Moscow, p.97.
10. Gadaskina I.D., Tolokontsev N.A. (1988) *Poisons - yesterday and today*, Nauka, Leningrad, p.176.
11. *World Ocean: Economics and Politics.* Edited by Academician Primakov E.M. (1986) Mysl', Moscow.
12. Nadtochiy G.L. (1985) *Geography of Marine Transport*, Transport, Moscow, p.199.
13. Feoktistov L.P. (1986) Armament Race and Scientific-Technological Progress are Incompatible, *Kommunist* **15**, p.99.
14. Balandin R.K., Bondarev L.G (1988) *Nature and Civilization* Mysl', Moscow, p.363.
15. Bochkareva T.V. (1988) *Ecological Threat of Urbanization* Mysl', Moscow, pp.3-40.
16. Kharitonov V.M. (1983) *Urbanization in the USA,* Moscow State University, Moscow, p.201.
17. Malyshev L.P. (1992), Military-ecological Analysis of Europe, in *Security-92,* General Staff of Russian Armed Forces, Moscow, p.73
18. Budyko M.I. and al. (1986), *Global Climatic Catastrophes,* Gidrometeoizdat, Moscow, p.86.
19. Aizatulin T.A. (1992) *Chemical Danger,* Fund of National and International Security, n.3 p.37.
20. Blinov A.Yu. (1992) *Airborne Subsystem for Environmental Monitoring and for Actions in Emergency Situation of Technogenic Catastrophes*" State Research Institute, p.38.
21. Mikhno E.P. (1979) *Elimination of the Consequences of Emergencies and Natural Calamities* Atomizdat, Moscow, p.19

22. Skryagin L.N (1986), *Secrets of Marine Catastrophes,* Transport, Moscow.
23. Marshall V. (1989), *Main Dangers of Chemical Productions,* Mir, Moscow.
24. Nikitin D.P., Novikov Yu.V. (1986) *Environment and Man,* Vyshaya shkola, Moscow, p.58.
25. Lemeshev M.Ya. (1991) *There is still Time,* Molodaya Gvardiya, Moscow.
26. Yablokov A.V. (1991) A Country - Mutant, *Ekos* **2**, p.36.
27. Malyshev L.P (1993), Though the Weapon is Conventional, *Green Cross.*
28. Malyshev L.P., Surikov B.T. (1993) Memorandum "*Armament Race is Extremely Dangerous for the Community of CIS countries*", RF Supreme Soviet, Moscow.
29. Malyshev L.P. and al., edited by Trubitsyn O.N (1993) *Report on Research Project "Klug": Research Investigation and Development of Unified Subsystems and Technologies for Creation of Robots of Marine Support,* Russian Academy of Sciences, Moscow, p.27.
30. Malyshev L.P. (1992) Is a High-accuracy Weapon an Alternative to Nuclear Weapon? *Aviation and Cosmonautics* **3-4**, p.10.
31. *Nuclear arms and Sovereignty of Republics* (1992) Prepared by Center for Disarmament and Strategic Stability, Mezhdunarodnye otnosheniya, Moscow.
32. Malyshev L.P. (1993) START-2 and Ecological Safety of Russia, *Green World* **11**, pp.1,7,10.
33. GAO: United States General Accounting Office. Report to the Acting Secretary of the Navy Nuclear Submarines. *Navy Efforts to Reduce Inactivation Costs.* July 1992 GAO/NSLADF - 92/134, United States General Accounting Office, Washington, DC 20548.
34. Litovkin B.N. (1993), 93 Nuclear - Powered Underwater Ships are Waiting Utilization in Russia, *Izvestiya*, July 10.
35. Kirilenko V.P., Malyshev L.P., Skaridov A.S. (1992) *Working out of the Principles of Law Guarantees of Ecological Safety of the Processes of Military Activity in all Stages of the Life Cycle of Weapon Systems and Technical Facilities* Fund of National and International Security, Moscow.
36. Problems of Chemical Weapons Elimination (1993) *Russian Chemical Journal* **3**
37. Report on research project "*Complex Analysis of the Danger from the Captured German Chemical Weapons Dumped in the Baltic Sea in the post-War period*", military unit 52688, research group under Dr. Kholostov V. Moscow, p.39.
38. Resolution 44/21 of the UN General Assembly: *Strengthening of the World Peace, Security and International Cooperation in all its Aspects in compliance with the UNO Charter,* New-York. November 15, 1989, UN, New York 1990, p.391.
39. Borisov T.N., Golosov R.A., Ivanitskiy L.N., Malyshev L.P., Slipchenko V.I., Shevelev E.G. (1992), *Basics of Methodology of Complex Investigation of Ecological Problems and Substantiation of the Methods to Ensure Ecological Safety of the Military Activity of Russia,* Fund of National and International Security, Moscow.
40. Kuznetsov G.I. and al., edited by Utyakov L.L. (1992) *Scientific-technical Concept of Long-term Ecological Monitoring of Aquatories of Enhanced Danger,* Russian Academy of Sciences, Moscow.

41. Malyshev L.P. and al., edited by Trubitsyn O.N. (1993) *Report on a research project "Klavessin-OV": Research Investigations for the Problem of Creation of Systems of Technical Vision for a Series of Underwater Robots with the Use of Elements of Artificial Intelligence,* Section of applied problems, Russian Academy of Sciences, Moscow.

APPLICATION OF ANTI-FILTERING COATINGS FOR LOCALISATION OF TOXIC WARFARE CHEMICALS IN THE BALTIC SEA AREA

V.G. PLOTNIKOV, R.A. ZAMYSLOV, B.T. SURIKOV,
I.V. DOBROV, O.Yu. KAYURIN
Obninsk Branch of the Karpov Institute of Physical Chemistry
Obninsk 249020, Russia

According to official reports, the period from January 11, 1946 to July 3, 1951 was remarkable for the submersion of chemical weapons in the Baltic Sea. The ammunition captured from Germany contained a great variety of chemical aerial bombs, mines, flue and chemical fougasses, shells, grenades, roll and drum-shaped containers of different capacity all filled with yperite, chloroacetophenone (CN), arsenic oil, adamsite, dichloroarsine. Officials say that a total amount of 300,000 tonnes of warfare chemicals were plunged into the sea water in more than 600,000 different objects. The items of chemical ammunition were either scattered loosely or dumped at the sea bottom loaded on board out-of-date vessels and barges. The majority of these dumping locations are well-known, but some of them remain uncertain because of the lack of appropriate information. Experts express a great deal of concern about the possible spreading of toxic products in the Baltic Sea in the near future, and predict disastrous consequences of beach pollution, mass intoxication in densely populated onshore areas, etc.

Three approaches can be suggested to prevent this ecological disaster:

1. To raise chemical-containing objects from the sea bottom with all preventive measures being taken, e.g. freezing to low temperatures and degassing. However, this approach requires considerable investment to construct freezing plants for the subsequent dismantling of explosive objects and war chemical degassing. Thus, this is not a perfect solution to the problem of the Baltic Sea environment's protection.

2. Degassing of war chemicals in near-the-bottom water layers. However, a problem arises due to the low temperatures of the bottom water layers and the consequently slows the rate of neutralisation processes. In addition, interlayer streams can dilute the concentration of reagents. As a result, the reaction will be incomplete and sea water pollution will increase considerably. Certain chemicals will decompose upon contact with sea water and the toxic products of their decomposition will be stream-drift farther.

A. V. Kaffka (ed.), Sea-Dumped Chemical Weapons: Aspects, Problems and Solutions, 105–107.

3. The most probable and dangerous consequences of chemical weapon decomposition arise from arsenic-containing compounds, i.e. lewisite, adamsite, arsenic oil, etc. If the source of arsenic-containing and other decomposition products is identified in a timely fashion, it is possible to prevent toxic migration by applying coatings based on water-swelling polymers. A number of polymers are known to be water-swelled and to slow down water filtration through the polymer layer or polymer-containing composition mixed with other materials of both organic and inorganic origin.

Application of water-swelling polymers to fasten soil onto the surface of the ecologically dangerous objects and to decrease the filtering capability of the soil seems to be a promising approach to prevent war chemical spreading in the Baltic Sea.

Antifiltering screens have been the subject of considerable research both in Russia and abroad. Antifiltering methods are based on the two following approaches:

- injection of low-viscosity monomer solutions into the soil to form a hard polymer screen of high protective efficiency;
- injection of water-swelling polymer pulp into the soil to create a swelled polymer screen against the diffusion of toxic products.

The top ground layer can also be protected by a layer of swelling polymers. Most suitable for this purpose are inorganic high-molecular-weight compounds of silicate solutions and their derivative, as well as cement-colloid solutions. Highly water-resistant polymers of the acrylate type seem to be most efficient in this specific application. It is also possible to use mixed cement-colloid solutions. Along with cement, successful results can be achieved with any filler, sorbent, lime or a special additive catalysing warfare agent decomposition and absorbing toxic products through absorption or ion exchange.

Hardening and insulating compositions must be ecologically safe and highly penetrative, with the time of hardening (gel formation) easily controllable. It is important that polymerisation time should be independent of temperature to enable the process completion in a temperature range of 0-15°C under rigid conditions of high salt content, to yield a hard polymer highly resistant to water, ageing, chemical, physical and biological effects.

To meet the requirements listed above, the polymers must be chosen among polyacryl amide or polyacrylic acid compositions, acrylic amide copolymers or a water solution of acrylic amide with cross-linking additives.

To monitor any release of warfare chemicals into the sea water, the dumping locations should be equipped with special automatic analysers to transmit timely radio signals to the bases.

We suggest to develop the monomer-polymeric composition based on acrylic amide and acrylate polymers with adsorbing fillers and CW-neutralising additives. Polymer content in the composition must not exceed 4%.

The suggested method will prevent or considerably slow down the speed of spreading of toxic products in the sea water. Polymer coatings based on polyacryl amide are nontoxic materials which are also highly resistant to adverse environmental effects. Each type of warfare agent requires the development of a specific polymer composition.

CHEMICAL-ANALYTICAL CONTROL OF ENVIRONMENTAL POLLUTION BY WARFARE AGENTS AND THEIR DEGRADATION PRODUCTS

Yu.I. SAVIN, E.M. VISHENKOVA,
E.M. PASYNKOVA, I.S. KHALIKOV
Institute of Experimental Meteorology
SPA "Typhoon", Roskomhydromet
Lenin str. 82
Obninsk, Kaluga region, 249020 Russia

1. Chemical-Analytical Control of Warfare Agents

The elimination of mass destruction weapons, including chemical weapons, is one of today's key problems. Enormous stocks of chemical weapons (CW) and the consequences of weapons production, testing, utilisation and destruction, among them dumping into the world's oceans, present a serious threat to the environment and human health. Therefore, the problem of ecological safety measures connected with CW destruction is rather urgent.

In connection with the realisation of the Program Of Phased CW Destruction In Russia, first it is necessary to qualify the environmental conditions in regions of chemical weapons storage, destruction and former production or, in other words, to carry out comprehensive studies of areas which have been related or are now relevant to chemical weapons, their pollution by chemical warfare agents (WA), their transformation products (TP) and another toxic compounds.

The second, not less important problem is to organise and to carry out permanent chemical-analytical control of WA and TP in the environment during the whole period of CW destruction.

Chemical analysis allows us to determine the polluted areas, to ascertain the dangerous substances and the degree of pollution, as well as to obtain information about the sources and the principle means of pollution.

Information about dangerous substance content in the environment in its turn is necessary to reveal the cause-effect relationships between environmental pollution and its ecosystem and public health impacts, and to make decisions about the prevention of dangerous substance intake in the environment and on polluted object purification.

Activities in estimating the degree of WA (and TP) environmental pollution will entail difficulties associated with inadequate information on the main patterns of WA migration and transformation in the environment, insufficient data on the toxicity and

A. V. Kaffka (ed.), Sea-Dumped Chemical Weapons: Aspects, Problems and Solutions, 109–118.

stability of some transformation products, the scarcity of physico-chemical methods to analyse WA and the toxic products of their transformation in the environment.

The organisation and realisation of chemical-analytical control of WA, their transformation products and substances formed as a result of technological processes of WA destruction and utilisation in the environment of indicated regions are impossible without solutions to a series of complex and interrelated problems concerning:

- scientific-methodological support;
- substantiation of a list of obtained data and parameters necessary to estimate the environmental conditions and to forecast a situation;
- selection and substantiation of a list of controlled substances and objects;
- methodological, meteorological and instrumental support;
- technical support.

1.1. SCIENTIFIC-METHODOLOGICAL SUPPORT

Information on the ways and mechanisms of WA migration and transformation in the environment, main transformation products, their lifetime in the environment and other characteristics of WA behaviour is extremely important for monitoring the environment in the regions of CW production, testing, storage and destruction.

Unfortunately, these data are scarce in literature and refer to the original combat warfare agents only. Data on toxic WA transformation products, formed both as a result of WA intake in the environment and technological processes associated with WA destruction and utilisation, are practically non-existent. However, proceeding from the physico-chemical properties of some warfare agents (e.g. yperite and lewisite), transformation products comparable with the original compounds in toxicity and sometimes exceeding them by 2-5 times are expected to be formed. In these cases, their stability in the environment is unknown.

That which was said above emphasises the need for fundamental research directed at establishing the main patterns of WA behaviour in the environment, toxic products of WA transformation and estimating their stability, and also developing or improving and certifying methods for their analysis, necessary to realise chemical-analytical environmental monitoring in regions relevant to warfare agents.

This research should be carried out in the following areas:

- selection of methods and development or improvement of techniques for instrumental WA and TP analysis to study the behaviour of these substances in the environment;
- determination of WA and TP degradation rate in soil, water and bottom sediments depending on temperature and acidity (pH);
- estimation of the redox-potential effect on WA and TP transformation rate in soil, water and bottom sediments;
- estimation of the effect of photochemical and microbiological processes on WA and TP disintegration rates;

- determination of WA and TP absorption characteristics in different soils and bottom sediments;
- estimation of WA and TP migration capacity in the following systems: soil-plants-herbivorous animals and bottom sediments-water-marine animals;
- establishment of the prevailing migration and transformation means of warfare agents and their derivatives in water and soil, and determination of the kinetic parameters of these processes for prognostic models of ecological safety in controlled regions.

1.2. A LIST OF DATA AND PARAMETERS TO BE OBTAINED

Concentrations of WA and their toxic transformation products, formed both as a result of technological processes associated with CW destruction and WA intake in the environment, should first be measured in regions of CW production, testing, storage and destruction as well as in their background areas.

Continuous or rather frequent measurements may be necessary to take account of the dynamics of toxic substance intake in the environment. Simultaneous measurements of several components in different media are needed to determine the relationship between the levels of hazardous substances in them and to establish the means of their migration and transformation.

Migration and transformation rates of WA and their toxic derivatives under various environmental conditions also need to be known. With this end in view, it is necessary to determine the rates of evaporation, solution, sedimentation, degradation periods and lifetimes, distribution coefficients and other characteristics of the substances' cycles in nature.

It should be noted that a part of the above parameters may be estimated in the laboratory.

1.3. CONTROLLED SUBSTANCES AND OBJECTS

A list of controlled substances will depend on their toxicity and stability in the environment, technological processes of destruction and utilisation and may be definitely determined only after preliminary laboratory studies on the primary means of WA migration and transformation and establishment of their main transformation products.

Concentrations of WA and their derivatives should be firstly measured in air, soil, surface, ground and discharge waters, bottom sediments, plant and animal substances as well as in industrial wastes. In some cases, WA and TP concentration measurements may be necessary in other objects potentially hazardous for the environment.

1.4. METHODICAL, METROLOGICAL AND INSTRUMENTAL SUPPORT

The next set of problems connected with monitoring hazardous substances in regions of the storage, destruction and former production of chemical weapons refers to analytical support. The state of the current situation is so, that now the controlling departments have no reliable methods to analyse WA and especially their toxic transformation products, no proper equipment or metrological support.

Warfare agents, which are subject to destruction, refer to the so-called group of supertoxicants including polychlorinated dibenzodioxins, dibenzofurans and biphenyls, organic phosphorous compounds, 3,4-benzopyrene, nitrosocompounds, etc. Analysis of these substances requires a high selectivity factor at low detection limits. The required detection limits of superecotoxicants in composite multicomponent mixtures are within 10^{-7} - 10^{-12} mg/l. A high selectivity factor in the analysis may be achieved by highly effective chromatography or highly informative analytical methods (chromato-mass-spectrometry or chromato-IR-Fourier-spectroscopy) with unique analytical equipment available only in a few Russian specialised centres. Such Russian-made equipment needs development, therefore it is preferable to purchase it abroad. The development or improvement of methods to analyse WA and TP in the environment, their metrological support and certification are also complex, time-consuming problems.

1.5. STANDARDS AND TECHNICAL SUPPORT

Until now, little emphasis was placed upon this question. At this time, only the maximum permissible concentrations of WA in air for working zones and settlements are established. For water and soil, the maximum permissible concentrations of WA are not established. Data on the maximum permissible level and the maximum permissible concentration for toxic WA transformation products is also lacking.

Relative safety standards regulating the maximum permissible concentration of warfare agents and, especially, their toxic transformation products in environment and living organisms either are not yet developed or are being developed extremely slowly.

Monitoring of WA and their TPs in water, air, soil, etc. is impossible without these standards.

The lack of information on technological rules, concerning industrial wastes in particular, and on the former production outputs of warfare agents does not allow us to estimate the amount of pollutant emissions into the environment.

We have initiated studies on WA behaviour in the environment within the scope of possible fields of works.

2. Yperite and Lewisite Behaviour in Water

2.1. DEPENDENCE OF YPERITE HYDROLYSIS RATE ON pH AND TEMPERATURE

Hydrolysis rates of dissolved yperite were studied in a phosphate buffer prepared with deionised water (R = 10^{-18} Mom x cm) at the temperature of 17 ± 1°C. Yperite was detected by highly effective liquid chromatography using an LKB liquid chromatograph (Sweden) with a UV absorption detector at a wavelength of 210 nm.

As seen in Table 1, the hydrolysis of dissolved yperite takes place extraordinarily quickly. The highest rate of hydrolysis was observed at pH 4.65 ($k = 0,095 \pm 0,003$ min^{-1}). In more acidic or subalkaline media yperite's hydrolysis is less intense. Yperite's hydrolysis is 4-5 times slower in sea water, and this should be taken into consideration when estimating the hazard of chemical weapons dumped in different areas of the world's ocean.

TABLE 1. Dependence of yperite hydrolysis rate on pH

pH	k, min^{-1}	T_{50}, min	T_{99}, min
2,65	0,066 ± 0,003	10,5	69,8
3,65	0,086 ± 0,003	8,1	53,6
4,65	0,095 ± 0,003	7,2	48,5
6,50	0,077 ± 0,002	9,0	59,8
6,90	0,071 ± 0,008	9,8	64,9
7,57	0,051 ± 0,004	13,6	91,3
6,80*	0,016 ± 0,005	43,3	288,0

* water sample from the Barents Sea (t=20°C)

Temperature effects on the rate of yperite's hydrolysis were similarly studied at temperatures of 9,14 and 24 °C.

TABLE 2. Temperature effects on yperite's hydrolysis rate

Temperature, t °C	k, min^{-1}	T_{50}, min	T_{99}, min
9	0,023± 0,002	30,0	200
14	0,042± 0,003	16,5	110
24	0,123± 0,007	5,6	38

As seen in Table 2, the environmental temperature greatly effects the rate of yperite's hydrolysis. With a 10 °C rise in temperature the hydrolysis rate increases approximately by 3 fold. At lower temperatures the hydrolysis rate decreases notably. Once in water with a lower temperature (2 °C), yperite "leaches" and settles to the bottom as a drop which has not managed to disperse or to dissolve during a 10-min stirring; therefore it may be assumed that at lower temperatures yperite can keep for a long time. This conclusion is extremely important when estimating the stability of dumped chemical weapons as, for example, the annual mean temperature at the bottom of the Baltic Sea is 2-4 °C.

The activation energy of yperite, calculated from the Arrenius formula, is 19.4 kcal/mol.

2.2. YPERITE'S HYDROLYSIS RATE IN A HETEROGENEOUS PHASE

Processes proceeding with yperite under heterogeneous conditions are more complex in comparison with homogeneous ones. In this case most of the yperite is not dissolved in water. Yperite dissolution goes much slower than the hydrolysis of its dissolved part, and therefore it is a limiting stage of the process.

The dissolution of yperite, ingressed in water, will strongly depend on the substance amount and the area of phase separation. So, according to the literature's data, a drop 1 cm in diameter was dissolved with a semidisappearance period of 15 days. According our data, a drop 6.4 mkl in volume was dissolved over six days The environmental temperature and the mixing intensity will also strongly affect the rate of yperite dissolution.

As the hydrolysis of dissolved yperite proceeds, a non-dissolved substance transforms into solution and is subjected to diffusion and hydrolysis. Compounds, formed as a result of hydrolysis, hinder further yperite dissolution and simultaneously interact with yperite and between themselves forming a series of toxic sulfonium compounds including 1,2-bis-(β-chloroethylthio)ethane which is five times more toxic than yperite itself. Hydrogen chloride, released as a result of yperite hydrolysis, may react with β-chloro-β-oxidiethyl sulphide and thiodiglycol and form the initial compound. With hydrogen chloride acceptors in the water the equilibrium of the reaction will displace to the right. Without them there may be an equilibrium which favours a long-term preservation of yperite in the aquatic medium.

Experiments with simultaneous observations of decrease in yperite's drop sizes and its concentrations in water have been performed to estimate the equilibrium concentrations of yperite and the duration of its preservation under heterogeneous conditions. It was also interesting to estimate the effects of micro-organisms on yperite's transformation rate.

For the experiments we have used natural water from the Protva river with pH 8.5, the same water which was sterilised by filtering it through a membrane filter with a pore diameter of 0.3 mkm and also deionised water with pH 7.2. Yperite was extracted with hexane and analysed by gas chromatography using a "Crystal - 2000" chromatograph model III with a PFD-S detector.

Investigation results have shown that in non-sterilised water the concentration of yperite in a solution quickly increases, reaches a maximum in one day and then decreases to a certain constant level (the end of drop dissolution) corresponding to the equilibrium concentration of yperite. Equilibrium was observed 4-5 days after the beginning of the experiments. Yperite concentrations in the water at equilibrium were 10-30 mg/ml (2-5% of the initial concentration). In deionised water the equilibrium concentration even after a year was about 2%. In sterilised natural water the curved pattern of measured yperite concentration was preserved. However, maximum yperite concentrations in water were approximately three fold higher than in the non-sterilised version. Equilibrium was observed after 6-7 days, in this case the yperite content was observed at the level of 3-6% of the initial content. The increase in yperite's transformation rate during the first period is possibly caused by micro-organism effects. However, final conclusions can be made only after additional studies. It is clear that these data are insufficient to estimate the period of yperite preservation under the appropriate natural conditions, but they may throw light on its behaviour.

So, the obtained results show that, in the case of yperite intake into water, it can keep for a long time (up to a year or longer) at high concentrations that give highly toxic properties to the water reservoirs.

2.3. DEPENDENCE OF YPERITE'S TRANSFORMATION RATE ON HYDROGEN PEROXIDE

The presence of two free electron pairs on a sulphur atom in the yperite molecule gives nucleophylic properties to yperite, or the capacity to undergo electrophylic reactions such as oxidation or chlorination. Oxidisers (hydrogen peroxide, chlorine, nitric acid, permanganates, chromic acid, hypochlorites of alkali and alkaline-earth metals, etc.) transform yperite into toxic β,β'-dichlorodiethylsulphoxide and β,β'-dichlorodiethylsulphone. In the abundance of oxidisers, the process continues through to the formation of β-chloroethynsulphonic acid which undergoes full destruction under more severe conditions.

As hydrogen peroxide is contained in natural water as a product of the vital functions of micro-organisms in concentrations of about 10^{-5} M, we have studied its effect on the rate of yperite oxidation.

As seen in Table 3, the rate of yperite transformation with no addition of hydrogen peroxide is caused by its interaction with water, i.e. hydrolysis.

TABLE 3. The effect of hydrogen peroxide content on yperite transformation rates (C_O=1,6*10^{-3} mol/l; t= 9°C).

Concentration of H_2O_2, mol/l	**Rate constant (k), min^{-1}**	**T_{50}, min**	**T_{99}, min**
0	0,023± 0,002	30	200
0,00001	0,022 ± 0,002	32	212
0,0001	0,036 ± 0,004	19	128
0,01	0,038 ± 0,005	18	121
0,1	0,046 ± 0,010	15	99

Hydrogen peroxide at $1*10^{-5}$ mol/l concentration increases the rate of yperite transformation by a factor of 1,5-2. The presence of stronger oxidisers in natural water may make a greater contribution to the total rate of yperite transformation.

2.4. BEHAVIOUR OF LEWISITE IN WATER

The literature data on Lucite's behaviour in the natural environment is scarce. There is only evidence that the hydrolysis of lewisite proceeds at a higher rate as compared to yperite. In this context we have attempted to estimate the rate of its hydrolysis by IR-Fourier spectrometry.

It turned out that lewisite reacts with water immediately (in fractions of a second) and forms a slightly soluble compound β-chlorovinylarsineoxide. In the case of large amounts of lewisite being ingressed in water, it will deposit on the bottom being immediately covered by a layer of β-chlorovinylarsineoxide, which prevents lewisite's dissolution and hydrolysis.

2.5. STABILITY OF β-CHLOROVINYLARSINEOXIDE IN WATER

The dependence of the rate of β-chlorovinylarsineoxide (a lewisite hydrolysis product) degradation on the pH of aqueous solutions was studied. Buffer solutions and deionised water with pH 6.4 were used. The content of β-chlorovinylarsineoxide was determined by an HLEC method. Experimental results are presented in Table 4.

As seen in Table 4, β-chlorovinylarsineoxide in aqueous solutions- is rather stable against degradation. Its stability increases in more acidic media. In contrast, in neutral and slightly alkaline media the rate of β-chlorovinylarsineoxide degradation is approximately two fold higher ($T_{1/2}$=20-25 days) than in acidic ones.

TABLE 4. The dependence of β-chlorovinylarsineoxide degradation rate in water on pH (t = 22 °C)

pH	k, day^{-1}	$T_{1/2}$, day	r
6.4*	0,011 ± 0,004	62	0.80
4.4	0.013 ± 0,003	53	0.90
5.8	0.018 ± 0,003	38	0.92
6.4	0.028 ± 0,012	25	0.84
9.4	0.033 ± 0,003	21	0.98

* deionised water without a buffer

The degradation of β-chlorovinylarsineoxide in natural water seems to proceed not only at the cost of hydrolysis, but also as a result of oxidation to β-chlorovinylarsonic acid which has no cutaneous effects.

So, from the above remarks it may be concluded that in aqueous solutions the product of lewisite decomposition (β-chlorovinylarsineoxide) may be preserved much longer than the initial compound and result in the long-term pollution of the natural environment.

Conclusions about the long-term preservation of toxic WA transformation products in the environment were confirmed by studies of the territory near Chapaevsk, where chemical weapons were produced from 1943 to 1945.

3. Results of Environmental Studies near Chapaevsk, Samara region, for Warfare Agents (Yperite, Lewisite) and Their Transformation Products, in 1993-1994

The studies were aimed at searching for the sources contaminating territories near Chapaevsk with warfare agents (yperite, lewisite) and their transformation products, and establishing the means of WA migration and transformation in the natural environment.

Snow, water and bottom sediment samples collected in 1993-1994 were analysed for their content of WA agents (yperite, lewisite) and their transformation products by the methods developed at SPA "Typhoon".

Considering the extremely low persistence of lewisite and the long time period after the stoppage of its production, only transformation products may be found in soil, e.g. β-chlorovinylarsineoxide (a sufficiently stable product of lewisite's hydrolysis), arsenic oxide as well as arsenates and arsenites. Due to this, the collected samples of soil, water and sediments were first analysed for their gross arsenic content.

The arsenic content exceeded maximum permissible concentration (M.P.C.) in practically all samples. In residential districts the arsenic content ranged from 4 to 10 M.P.C. The territory of Chapaevsk Chemical Fertilise Plant (CCFP) was greatly polluted by arsenic. The arsenic content in samples collected in this territory exceeded 15 M.P.C., and extremely high concentrations of arsenic (from 49 to 8500 M.P.C.) were registered in samples collected in the former lewisite-producing shop and near the place of WA bottling.

The analysis of bottom sediments from the Chapaevka river also had shown an excess of arsenic content, i.e. from 2 to 17 M.P.C. The arsenic content in the drinking water and snow turned out to be lower than the maximum permissible concentration.

The found high concentrations of arsenic may be caused both by inorganic and organic arsenous compounds and α-lewisite or its hydrolysis product β-chlorovinylarsineoxide in particular.

In the natural environment, α-lewisite quickly interacts with water, forming toxic β-chlorovinylarsineoxide. It is a sufficiently stable and water-insoluble compound which also involves β-chlorovinyl group. So, it was supposed that α-lewisite in the environment (water, soil) could persist for as long a time as β-chlorovinylarsineoxide. It was shown that when interacting with alkali, β-chlorovinylarsineoxide decomposes as a transisomer of α-lewisite with release of acetylene. Therefore, to detect α-lewisite and its transformation product β-chlorovinylarsineoxide in soil we have used the method of detecting acetylene released after applying alkali to the sample, with the help of gas chromatography with a flame-ionisation detector (FID)

All soil, water and sediment samples collected in Chapaevsk were analysed by the acetylene method. Acetylene release was observed in 9 samples.

To prove the fact that the acetylene formed in an alkali treatment of the samples was caused by the β-chlorovinylarsineoxide present in them, we have identified it with the help of chromato-mass-spectrometry in the form of diethyl ether of chlorovinylarsineoxide, using a Finnigan MAT trap with ITD-800 and a 30 m x 0.32 mkm HP-I column.

β-chlorovinylarsineoxide was extracted from the soil samples and transformed to diethyl ether using ethyl alcohol. The obtained extracts were quantitatively analysed in a Crystal-2000 gas chromatograph with ESD.

Chromato-mass-spectrometer analysis of the diethyl ether from β-chlorovinylarsineoxide in the soil and sediment samples has confirmed that acetylene formed in alkali sample treatment is caused by the β-chlorovinylarsineoxide observed in them. The greatest amount of β-chlorovinylarsineoxide (from 2 to 32 mkg/g of soil) was detected in samples collected in the territory of CCFP (former lewisite-producing factory), as well as in waste sediments from CCFP.

Extraction with organic solvents was used to extract yperite and its transformation products from the soil. Hexane was used successfully (up to 80%). for extracting yperite from soil samples. Yperite transformation products (β, β'-dichlorodiethylsulfoxide, β, β'-dichlorodiethylsulfone and thiodiglycole) were extracted from the soil with acetone. Acetone extraction of these compounds from soil samples amounted to 30% for thiodiglycole, 70% for β, β'-dichlorodiethylsulfoxide, 75% for β, β'-dichlorodiethylsulfone.

Chromatograms and mass-spectra of the reference yperite samples and its transformation products were preliminary obtained in the gas chromatograph and ion trap. The residence time of these compounds was: 5.07 min. for yperite, 5.21 min. for thiodeglycole, 7.67 min. for β, β'-dichlorodiethylsulfoxide, and 7.87 min for β, β'-dichlorodiethylsulfone.

All samples collected in Chapaevsk were analysed for their content of yperite and its transformation products. Yperite and its transformation products were not detected in samples with the help of the described methods. However, in samples collected in the territory of CCFP, a great variety of organic substances were found, among them HCCH, hexachlorobenzol, DDT, benzapyrene, etc.

It should be noted that the estimation of environmental contamination caused by warfare agents (yperite, lewisite) and their transformation products near Chapaevsk, according to the "Criteria for estimating ecological situation to establish regions of extreme ecological condition and ecological disaster" (an instruction issued by the Ministry of Environment Protection of Russia, 1992), was hampered by the lack of regular monitoring of environmental contamination by these substances. Because of difficulties and its high cost, the monitoring was of a sampling and episodic character. The high arsenic levels observed in the territory of CCFP, and its M.P.C. excess in the residential districts of the town are dangerous for the population; the situation is also aggravated by the fact that arsenic is partially in the form of a toxic product of lewisite's hydrolysis - β-chlorovinylarsineoxide - detected 50 years after the stoppage of warfare agents' production.

So, the solution to the complex problems associated with the organisation and realisation of chemical-analytical control of warfare agents and their transformation products in the regions of production, testing, storage and destruction is impossible without fundamental studies of their behaviour in the natural environment. Work in this direction is to be continued.

Section 2.3

Chemical Aspects

CHEMICAL WEAPONS ON THE SEABED

PROFESSOR G.V. LISICHKIN
Faculty of Chemistry, Moscow State University
Moscow, Russia

1. None Will Be the Wiser

Contamination of the world's oceans began to be considered an ecological problem of great concern only in the seventies. For the most part, this was connected with tanker catastrophes and petroleum pollution of the sea. Earlier, no one believed that the use of areas of water as a place for the burial of needless things was wrong.

That is why, it is little wonder that soon after the end of the World War II, leaders of the countries belonging to the anti-Hitlerite coalition made a decision to sink into the sea the captured chemical weapons of fascist Germany.

I think that we would do the allied war command an injustice, if we accuse them of such a barbarous decision, from the viewpoint of a present-day man. Indeed, it was necessary to annihilate the chemical weapons as soon as possible, lest anybody felt a wish to use them. At that time, however, methods of destroying contaminants in chemical plants with the observance of rules of safety procedures were not available. Even now, half a century later, there are no developed and completely reliable broad-scale technologies for the industrial destruction of war contaminants. Both concealment in deep mines and combustion of contaminants have more harmful effects on nature and mankind when compared to sinking these at a considerable depth. One important point to remember is also that German depots of chemical weapons were located in densely populated regions of Europe; therefore, the chosen method for getting rid of chemical ammunition was apparently the best.

2. What Is Found on the Seabed?

Captured chemical weapons comprised hundreds of thousands of tonnes of various contaminants: yperite and its derivatives, arsenic-containing compounds (lewisite and so on), chlorine-containing substances (phosgene), hydrocyanic acid, and organophosphorous compounds (tabun). The weapons were stored (both as ready ammunition - aerial bombs, shells, and mines - and in barrels and kegs) in several arsenals in occupied Germany.

A. V. Kaffka (ed.), Sea-Dumped Chemical Weapons: Aspects, Problems and Solutions, 121–127.

A decision was made, according to which the war administrations of each occupied zone had to independently destroy those chemical weapons located in its territory. The greatest quantity of weapons was revealed in the territory occupied by British forces. In the Soviet zone, about 40 thousand tonnes of war contaminants was found.

It was suggested that chemical ammunition would be sunk into the Atlantic Ocean at a depth of no less than 1,000 m. Unfortunately, the Allies had no available or reliable means of transport capable of a long voyage with tens of thousands of tonnes of war contaminants aboard the ships. Therefore, managers of the operation reduced the requirements for the depth of sinking, decreasing its value to 100 m. This allowed them to restrict the region of sinking to the Baltic Sea. The Englishmen charged chemical ammunition into the holds of old ships, cemented the decks, and sank the ships in the Straits of Belt Minor and Skagerrak near the Norwegian coast at a depth of 600 m. The Soviet part of the chemical trophies was sunk in bulk in two Baltic regions - near the Bornholm Island and between Liepaja and Gotland Island (Sweden).

In recent years, an assumption was made that chemical weapons had also been sunk in other areas - the Bay of Biscay, the White Sea, the Black Sea, the Japan Sea, the Arctic Ocean, and so on. Moreover, some authors state that until 1989, the USSR buried its own unneeded chemical weapons in the Baltic Sea [1] However, no official confirmations of such facts are available. In all publications, Russian military chemists report solely on the sinking of captured German chemical weapons into the Baltic Sea.

3. Contaminants in a Sea Medium

Considerable attention has recently been focused on the problem of rendering harmless and destroying chemical weapons. Clearly, the above subject has attracted considerable interest of the public along with that of specialists, because it is quite important to publicise key decisions in this field.

In a series of articles that appear periodically in newspapers and magazines, sunken chemical weapons are considered a delayed-action ecological mine of a sort, which will blow up in several months. In fact, if thousands of chemical bombs and shells were simultaneously destroyed owing to corrosion, and hundreds of thousands of tonnes of war contaminants got into the sea water and then into the atmosphere, nobody would remain alive over a vast area. On the other hand, more moderate predictions are published as well. However, in any case, sunken chemical weapons present a real threat for those populations which reside in the Baltic Sea region (about 50 million).

Let us try to consider the problem without bias.

First and foremost, one should take into account that the sea medium - precisely a medium and not merely water - imposes an extreme aggressiveness towards foreign objects, for example, a ship bottom, a manger board, or potato peelings that were thrown overboard from a liner. The following are the basic factors of the sea medium's actions: strong chemical and electrochemical corrosion, chemical oxidation of substrates by oxygen dissolved in water, microbiological oxidation and destruction, hydrolysis, photooxidation under the effect of the sun's radiation, and finally, the action of hydrobionts.

Contaminants themselves are also rather reactive - otherwise, they would have no effect on living organisms. Therefore, when in contact with an aggressive sea medium, contaminants begin inevitably to transform, yielding successively simpler and finally nontoxic compounds. This raises the question of the rate at which such a transformation occurs. If poisons are destroyed slowly, they are capable of wiping out a multitude of hydrobionts and then (via nutritive connections) inhabitants of dry land as well.

It is obvious that the action of chemical weapons results in the most terrible consequences in shelf and shallow water, where the majority of sea inhabitants are localised. It is also apparent that the consequences will mainly depend on the initial local concentration of contaminants in the sea water and on the currents, which carry around and dilute poisons. Concurrent release of a great amount of contaminants in a shallow inland sea, where water circulation is rather slow, would result in a catastrophe.

Unfortunately, the Baltic Sea is one of such inland seas, its depths being predominantly only 40 - 100 m. In addition, the Baltic Sea is subjected to considerable man-made pollution. We may have regrets that after World War II the victors over fascism failed to get the means to convey transports with chemical ammunition to deep-seated breaks in the Atlantic Ocean.

4. Catastrophe Schemes

On the Baltic seabed, the corrosion of contaminant-containing reservoirs has been proceeding for half a century. Clearly, the duration of corrosive destruction is different for thin-walled barrels, kegs with thicker walls, and for the heavy-walled bodies of aerial bombs. One should also take into account the differences in the grades of metals and in the location of the ammunition: some of them were found under a layer of sediments, others laid on silt ground and sank partially into it, and the third group were located on a solid bed and exposed to permanent currents, which accelerate corrosion.

According to an evaluation by academicians A.D. Kuntsevich, I.B. Evstafyev and their co-workers, 100% of barrels, 2% of kegs and certain individual aerial bombs should have already undergone a loss of sealing as a result of corrosion. Hence, it follows that the simultaneous seal failure of all reservoirs with contaminants is impossible.

What has happened when the seal failure of such a container laying at a seabed occurs? If the contaminant is liquid, its diffusion from the container into the sea water and the diffusion of sea water in the opposite direction begin to proceed through corrosion cracks formed, which are rather narrow early in the process. In the initial stage, the rate of contaminant outflow will be low, and the poison will be hydrolysed in a great bulk of water. If an eaten up keg is washed by a current, a poison will flow out and be destroyed at a relatively high rate. Thus, yperite, which is poorly soluble in water, is 99% converted into low-toxic diglycol at 10°C over 5 hours:

$$ClCH_2CH_2SCH_2CH_2Cl \rightarrow HOCH_2CH_2SCH_2CH_2OH.$$

The rate of hydrolysis of water-soluble organophosphorous compounds is even higher.

Until now, discussion has centred around liquid contaminants. However, poisons are often kept within ammunition as gels. Thickening agents were initially added to many yperite prescriptions, and yperite by itself is thickening gradually owing to polycondensation. When the corrosion destruction of a container filled with a high-viscosity or jelly-like contaminant proceeds, pollution of a sea medium occurs to a lesser extent, because a basic part of the poison remains inside the shell to the point of its complete destruction. Water penetration into a coagulate results in the formation of less toxic and less mobile oligomers in a contaminant bulk, thus further stabilising it.

True enough, such coagulates also contain a certain amount of the dissolved monomeric form of the contaminant. Beginning in the sixties, water from the Baltic Sea has been washing ashore chunks of toxic gel. Sometimes, such gel gets into the trawls of fishermen, but fortunately, no deaths have occurred to the present day. Nevertheless, prudent Danes built an installation (Bornholm Island) for the destruction of contaminant remains.

One can believe that a considerable part of the yperite and its derivatives currently in the ammunition will be found on the seabed as gel spots and solid deposits after the corrosion failure of the shells. As such spots and deposits become covered with ground sediments, the rate of contaminant hydrolysis decreases, but micro-organisms begin to work. A great number of microbe species are known, which can feed on substances that are toxic to a man. In 1991, NATO held a scientific conference devoted to biological degassing contaminants. In one of the reports there was mention of the detection of two species of pseudomonades in the ground silt of the Gulf of Mexico, which are capable of completely decomposing yperite.

Thus, hydrolysis, polycondensation, and microbiological destruction are those processes, which are able to render harmless sunken chemical weapons under natural conditions without the participation of man. It is thought that exactly the above considerations (along with the absence of cases of serious poisoning) are the reason for the rather indifferent attitude of the governments and the public of the developed Baltic states - Sweden, Norway, Denmark, Finland, and Germany - to the problem of sunken ammunition.

To be fair, it ought to be noted that specialists of the above countries have already done much to elucidate the situation. Thus, the Finns have collected a considerable amount of information on Baltic Sea contamination and on methods of the analysis of a sea medium. As a result of this work, a basic handbook in many volumes has been published. Norwegians performed sonar scanning of the seabed in a strait, and determined the state of ships sunken with chemical ammunition. The Swedes carried out a great many chemical analyses of the sea water and sea organisms in several places of contaminant burial.

Unfortunately, the situation provides well-founded reasons for alarm.

Let us imagine that the instantaneous seal failure of at least one out of thousands of large containers (filled with, for example, pure yperite or tabun) will occur owing to any accidental reason, for instance, as the result of a strong blow or an explosion. Experience shows that such contaminants, containing stabiliser additives, can be kept unchanged in a hermetically sealed tare for decades.

Recently, Professor S.S. Yufit analysed the effect of such a fairly realistic event. It follows from his data that if 100 kg of tabun got at once into the sea water, then with the rate of a bottom-current of 1 - 5 m/min, a 100 m thick water layer over the area of 100 km^2 will be severely contaminated. As this takes place, the contaminant concentration will be several orders of magnitude higher than the intolerable one in the close vicinity of the place of the seal failure. In the initial period, this concentration will exceed substantially the LD_{100} value (the concentration, at which 100% of experimental animals perishes). Of course, these values are of approximate character, but it is unlikely that exact calculations will lead to qualitatively different results.

The situation for contaminants, containing arsenic (lewisite, adamsite, and clarke-1) is distinct from the above one, but is not any better. The products of their hydrolysis are poorly water-soluble, but, what is more important, they are highly toxic. Therefore, if such contaminants get into a sea medium from the places of their burial, large areas of the seabed will be poisoned by highly toxic substances, in particular, by inorganic compounds of trivalent arsenic. Life will be impossible in these regions.

Thus, it may be concluded that passively waiting for an instant when there will be a crash of thunder is not the best strategy.

5. To Lift or Not to Lift

Do the above considerations mean that before the catastrophe occurs, sunken chemical weapons ought to be lifted and destroyed for the purpose of freeing the Baltic Sea and, possibly, other areas of water from contaminants once and for all? This is exactly what some specialists propose. However, let us try to estimate what needs to be done to achieve these results.

First, it will be necessary to dispatch no less than ten expeditionary ships of various classes for a detailed exploration of the seabed. These ships ought to be equipped with the most refined navigational systems, which allow one to plot on a map all dangerous objects, in particular, separately laying shells and bombs, with an accuracy of several meters. In addition, shipboard instruments should be used, which are able to detect sunken ammunition laying not only on the seabed , but under a layer of ground sediments as well. As a result of this stage of work, a detailed map of the Baltic Sea should be obtained, which shows virtually all chemical aerial bombs, shells, mines, containers filled with contaminants, and so on. However, the manner in which a man aboard an expeditionary ship can get information on certain critical questions is not clear enough (for example, is a container filled with a contaminant or is it a machine-oil tare, which was thrown overboard by a careless mechanic; what is the degree of the container failure due to corrosion; in what state - liquid or coagulated - is the contaminant?). Even though gas chromatographes are analysing water withdrawn by bathometers round the clock, we shall be able to get the exact answers to such questions only in rare instances.

The next stage of the project consists in fitting those ships intended for raising ammunition with equipment that is able to remove corroded containers with contaminants from the seabed surface, or more likely, from under a sediment layer. As this takes place, the possibility of spilling a contaminant and of contacting seamen with it should be completely prevented. Note that the instant of lifting contaminants is also of key importance in the context of ecological safety. Any accident should be ruled out, as the seal failure of a contaminant-containing reservoir during its raising is the evil against which the considered project is directed. In other words, 100% reliability in the lifting procedure is to be achieved. But is such a situation possible in reality, and especially in the sea?

So far, equipment for the safe lifting of chemical ammunition is not available, but interesting ideas connected with the use of a freezing technique are known. However, it is well to bear in mind that such equipment cannot be uniform, because both individual shells or kegs and barges of size 40 m are to be raised.

Now let us try to estimate, how much time the above procedures will require using the example of lifting sunken ammunition near Bornholm Island. According to the archive data, there are no less than 35,000 objects in this region. If we suggest that lifting a shell or a bomb takes two hours, then an operating ship would have to work continuously for eight years to complete the lifting.

Assume that we have succeeded in taking the ammunition on board without any accident. By this time, storages and chemical plants should be built on dry land in order to take contaminants from the ships and to destroy these, complying with the rules of safety procedures. Here also, complications inevitably arise.

The chemical plants for CW destruction are most likely to be constructed near St. Petersburg and Kaliningrad ports. I am not sure that the population of these regions will regard this idea with enthusiasm. In addition, the onset of the technological chain for contaminant treatment will be quite unusual, because a plant ought to be ready to take non-standard starting materials - all kinds of contaminants in any state in destroyed reservoirs.

Generally speaking, the modern-day state of science and technology allows us, in principle, to realise the above operations. However, if we wish to reach a high level of safety (of course, 100% safety is impossible), the costs of the operation will be very high - no less than $100 billion, and the operation will last, apparently, for about 15 years.

In summary, the lifting of chemical weapons is not to be carried out. A key argument for me consists in my personal experience participating in several projects connected with sea, fleet, chemistry, and complex technical equipment (although not in the high cost of the work and its considerable duration). This experience points unambiguously to a colossal risk, which the realisation of the idea of lifting chemical weapons involves.

Do the above considerations mean that I consider the problem of sunken chemical weapons to be hopeless, and we may only have hope in fate? No, they do not. Other non-dangerous and less expensive methods are available, which are capable of substantially decreasing the threat of Baltic Sea contamination. With this aim in view, we have to help the sea itself in destruction of foreign substances as soon as possible.

One such possibility consists in the acceleration of contaminant hydrolysis. Hydrolytic detoxification in an alkaline medium is known to be several orders of magnitude faster than that in a neutral medium. Therefore, one can cover underwater places of contaminant burial with any solid alkaline reagent, which will create an alkaline medium in the zone of the contaminant's location. Requirements for such a reagent are rather simple. It should be very poorly soluble in sea water, have a considerable specific surface area (no less than 50 m^2/g), display clearly defined alkaline properties and be a nontoxic, low-priced, and easily available substance. I believe that chemists-technologists working in the fields of silicate industry, metallurgy, and absorption processes already understand that a wide variety of well-known materials, including some industrial wastes, comply with the above requirements.

Thus, in my opinion, a constructive plan of action to save the Baltic Sea should be as follows.

First, extensive care must be taken to exclude the danger of an accidental unsealing in contaminant-containing places. In the regions where contaminants were sunk, the use of ground trawls, standing at anchor, geological explorations of all kinds, underwater explosions and laying submarines on the ground should certainly be forbidden by an international convention.

Then, all available archive materials on sunken chemical weapons, including co-ordinates, dates of sinking, kinds and amounts of ammunition, makes of contaminants, and their state at the instant the contaminants were sunk, should be declassified and published.

Thereafter, it is necessary to organise expeditions for the inspection of sites of sunken contaminants in order to create detailed maps of these places and neighbouring regions within 1 - 2 years. As we do not intend to lift each ammunition, such maps need not be as detailed as the ones mentioned above. Concurrent with these steps, prescriptions for active compositions, which can be used for chemical neutralisation and insulation of the ammunition, are to be developed on the basis of laboratory tests and full-scale trials.

And finally, upon the completion of all preliminary stages, the places of chemical weapons burial should be covered with selected reagents. This procedure appears to take 1- 2 years.

Thus, we have a good chance of saving ourselves and our descendants from the dangerous legacy of war. The cost of these works will be about $100 million. The ten countries of the Baltic Sea region can easily collect such a sum.

Reference

1. *Khimiya i Zhizn' (Chemistry and Life)* 7, 1993.

REACTION PRODUCTS OF CHEMICAL AGENTS BY THERMODYNAMIC CALCULATIONS

DR. F. VOLK
Fraunhofer Institut für Chemische Technologie (ICT)
D-76327 Pfinztal, Germany
Tel. 0049-721-4640-164
Fax 0049-721-4640-111

Abstract

In order to understand better the reaction behaviour of chemical agents of sea-dumped ammunition by thermal destruction in incinerators and plasma arc reactors, thermodynamic calculations have been performed. For this reason the following substances have been calculated at 1500 K:

- Clark I, Clark II, Adamsite and Lewisite
- Mustard such as S-Lost and N-Lost
- Tear gases: Chloroacetophenone and Bromoacetone
- Phosgene, Diphosgene and Chloropicrin
- Nerve gases Tabun, Sarin, Soman

A temperature of 1500 K was applied for the calculations because a plasma arc reactor is being used in Germany for the treatment of soil, contaminated with warfare agents, such as Clark I, Clark II, Adamsite, mustard, and tear gases.

The thermodynamic calculations have been conducted not only for stoechiometric combustion conditions, but also for a negative oxygen balance, which is relevant for partial combustion conditions and for thermal decomposition reactions. In this case, also the behaviour of chemical agents in pyrolysis processes can be determined.

By using an excess of water, also reaction products of a supercritical water oxidation can be forecast.

A. V. Kaffka (ed.), Sea-Dumped Chemical Weapons: Aspects, Problems and Solutions, 129–143.

1. Introduction

At the end of World War II, more than 300,000 tonnes of chemical weapon (CW) munitions were dumped in the Baltic and the North Sea [1]. This could pose a serious environmental danger to the security of the European nations. Therefore urgent steps should be undertaken to solve this environmental problem by elaborating appropriate technologies which are capable of decontaminating warfare agents.

Incineration is a widely used and proven technology with a long history of research and development. In the US incinerators are being used as deactivation furnaces for the treatment of ammunition containing gun propellants, high explosives, rocket propellants, igniters and primary explosives. It can be used also for remediation of soil contaminated with explosives.

In Germany, arsenicals such as Lewisite, Adamsite, Clark I and II, Sulphur- and Nitrogen mustards are destroyed by a special incineration method. Agents are drained into polyethylene barrels and incinerated directly at 1000°C [2]. Arsenicals are precipitated out of the incinerator scrubber water as $FeAsO_4$. The arsenic sludge is buried in drums in a waste dump (former salt mine). Since operation began in 1980, 73 tonnes of concentrated chemical weapon agents and 600 tonnes of contaminated materials have been destroyed. The cost estimate for Sulphur mustard is about 100 DM/kg.

2. Temperature Treatment

2. 1. HIGH TEMPERATURE TREATMENT

High temperature treatment requires consideration of reaction kinetics, decomposition kinetics and thermodynamics. These includes the rates of mass and heat transfer as well as chemical reaction rates [3].

Bimolecular processes such as the treatment with oxygen or with hot reaction products also contribute to destruction of organic compounds. The low activation energy processes are the reaction with radicals.

For example OH and H radicals are important in incineration processes. H radicals are important in pyrolysis, ions and electrons are present in plasma processing. For many compounds, bimolecular processes are the primary mode for destruction. If, however, the target molecule is thermal labile, it will decompose at low temperature without attack from radicals.

At the time being, a special incinerator is in construction for the treatment of soil contaminated with arsenicals. After a flotation process, soil wash containing the arsenicals goes to a plasma arc reactor where agents, carbon etc. are combusted and the soil is melted to a glassy slag containing some arsenic.

We assume that in the future another process will be of interest, namely the medium temperature treatment.

2.2. MEDIUM TEMPERATURE TREATMENT: SUPERCRITICAL WATER OXIDATION (SCWO) AND WET AIR OXIDATION (WAO)

These methods, also called hydrothermal processing [4,5], refer to oxidative reaction of wastes in water at elevated temperatures.

Supercritical reactions occur above the mixture critical point of about 500°C and 250 bar, so that the mixture is single phase. Wet air oxidation is carried out below the critical point. Commercial SCWO reactors with capacities of 400 L/day are available and a WAO in-ground system is now in testing [6].

3. Reaction Products of Chemical Agents by Thermodynamic Calculations

In order to understand better the reaction behaviour of chemical agents during the thermal destruction in incinerators and plasma arc reactors, thermodynamic calculations have been conducted at a temperature of 400°C in order to get the reaction products after the high temperature treatment.

The calculations have been done not only for stoechiometric combustion conditions, but also for negative oxygen balances, which are relevant for partial combustion conditions and for thermal decomposition reactions. By taking into account an excess of water, also the reaction products of a supercritical water oxidation can be calculated.

In this connection the following products have been calculated:

- Clark I, Diphenylchloroarsine (DA)
- Clark II, Diphenylcyanoarsine (DC)
- Lewisite, 2-Chlorovinyldichloroarsine (Agent L)
- Adamsite, Diphenylaminechloroarsine (DM)
- S-Lost, (Mustard Gas), Bis-(2-chloroethyl)sulfide (HD)
- N-Lost, (N-Mustard), N-ethyl-2,2'-dichlorodiethylamine (HN1)
- 2-Chloroacetophenone, (CN)
- Bromoacetone
- Chloropicrin, Nitro-trichloromethane
- Phosgene, Carbonylchloride (CG)
- Diphosgene, Trichloromethylchloroformate (DP)
- Sarin, Isopropyl-methylphosphonofluoridate (GB)
- Soman, Pinacolyl-methylphosphonofluoridate (GD)

- Tabun, Ethyl-phosphorodimethylamidocyanidate (GA).
- In addition, 2,4,6-Trinitrotoluene (TNT) was calculated.

4. Results

For calculating the products after reaction with air at one bar and with water at 220 bar - in both cases a temperature of 400°C was taken into account - the ICT-Thermodynamic Code was used [8].

In Table 1 (*see Tables and Figures in the Annex*) we see the results of the oxidation of Clark I and Clark II with air. In order to get also information about a partial oxidation of the warfare agents, in addition to the stoichiometric reaction with air (oxygen balance = 0), also a calculation for a negative oxygen balance at -30 g O_2 / 100g was carried out. By comparing the results we see a decrease of CO_2 and arsenic oxide and an increase of H_2, CH_4, of elemental arsenic and of solid carbon.

On the other side, the physical data [9] of the pure substances as well as the structural formula are given in Figures 1 and 2 (*see Annex*) for Clark I and Clark II and for Lewisite and Adamsite in Fig. 3 and 4. Because of the low carbon content of Lewisite, the oxygen balance of this substance is much less negative (-7.72 g O_2 / 100g) than that of Adamsite (-34.01). This means, the demand of air for a complete oxidation is much less than in the case of Adamsite, see Table 2. The reaction products with water at 400°C and 220 bar, which correspond to an oxidative reaction of warfare agents under supercritical conditions, are also presented in this table. As we recognise, the main reaction products are CO_2, H_2, CH_4 and HCl beside As and residual H_2O.

In Table 3, the reaction products of S-Lost and N-Lost are described. We see that the sulphur containing substance is going to form mainly H_2S; the possibility of SO_2 as a reaction product will increase with a better oxygen balance, this means with an excess of oxygen. On the other hand, chlorine has reacted completely to HCl. In connection with the physical behaviour we should notice that both substances exhibit very low melting points, see Fig. 5 and 6.

In Table 4, the reaction products of 2-Chloroecetonphenone, and Bromoacetone are contained. We see that chlorine reacts in each case to hydrogen chloride (HCl) and Br to hydrogen bromide (HBr). Other main products are CO_2, H_2O and Carbon soot. In addition, small amounts of H_2, CH_4 and CO are formed; see also Fig. 7 and 8.

In contrast with the last warfare agents, which contain enough hydrogen for the complete formation of HCl and HBr, chloropicrin, phosgene and diphosgene are without hydrogen, see Fig. 9-11 and Table 5. Therefore the formation of chlorine (Cl_2) must be taken into account. In addition, other chlorine containing molecules such as carbon tetrachloride (CCl_4) are possibly formed. In the case of phosgene and diphosgene, also carbon soot will be produced. But this carbon will be oxidised to CO_2 in the case of an excess of air.

Table 6 contains the reaction products of the warfare agents sarin, soman and tabun. Sarin and soman contain, in addition to carbon, hydrogen and oxygen also fluorine and phosphorus, tabun only phosphorus. Therefore reaction products such as HF, P_2O_3 and P_2O_5 will be formed, see Fig. 12, 13 and 14.

In order to be able to compare the warfare agents with a typical high explosive, Table 7 contains the reaction products of 2,4,6-Trinitrotoluene (TNT). As we see, reaction products such as CO_2, CO, CH_4, H_2O, N_2 and solid carbon are formed usually. Only in the case of a stoichiometric combustion or after the reaction of enough H_2O vapour, carbon dioxide and CH_4 are the products containing all carbon. See also Fig. 15.

References

1. Stock, Thomas, (1995) „*Scientific and technical aspects of sea-dumped chemical munitions*", Paper at the NATO Advanced Research Workshop on Sea-Dumped Chemical Weapons, Kaliningrad / Moscow, January 12-15.
2. Martens, Hermann (1994) „*Incineration of chemical weapons in Germany*", Paper at the NATO-Advanced Research Workshop on Destruction of Military Toxic Waste, Naaldwijk, The Netherlands, May 22-27.
3. Tsang, W. (1994) „*Incineration, pyrolysis and plasma arcs for destruction of toxic waste at high temperatures*", Paper at the NATO-Advanced Research Workshop on Destruction of Military Toxic Waste, Naaldwijk, The Netherlands, May 22-27.
4. Los Alamos National Laboratories, Los Alamos, NM 87545, USA.
5. Sandia National Laboratories, Livermore, CA 94550, USA.
6. „*Aqueous phase oxidation*", Ver Tech Treatment Systems, P.O. Box 292, 3740 AG Baarn, The Netherlands.
7. Hirth, Th., Eisenreich, N., Krause, H., Bunte, G., (1992) „*Entsorgung von Treib- und Explosivstoffen durch Prozesse in überkritischem Wasser*", Paper at the 23rd Int. Annual Conference of ICT, Karlsruhe.
8. Bathelt, Helmut and Volk, Fred, „*User's Manual for the ICT-Thermodynamic Code: Vol. 1, 2, 3*", Report 14/1988; 1/1991; 2/1991, Fraunhofer Institut für Chemische Techologie, 76327 Pfinztal, Germany.

Annexe

TABLE 1. Clark I and Clark II

Substance	CLARK I		CLARK II	
	with air (1 bar)		with air (1 bar)	
O_2-balance g O_2 /100g	0.	-30.	0.	-30.
Mole number gases (mol/kg)	31.77	29.25	32.00	29.71
Volume of gases (Nm^3/kg)	0.712	0.656	0.171	0.666
Composition (mol%)				
CO_2	16.18	10.35	16.27	10.58
H_2O	6.17	7.73	6.26	7.95
N_2	75.80	53.67	76.85	56.15
CO	-	0.20	-	0.21
H_2	-	2.02	-	2.06
CH_4	-	1.00	-	1.03
HCl	1.15	2.61	-	-
C (solid)	-	19.80	-	19.60
As	-	2.61	-	2.42
$AsCl_3$	0.07	-	-	-
As_2O_3	0.64	-	0.63	-

TABLE 2. Lewisite and Adamsite

Substance	LEWISITE			ADAMSITE		
	with air (1 bar)		80% H_2O (220 bar)	with air (1 bar)		80% H_2O (220 bar)
O_2-balance	0.	-30.	-7.72	0.	-30.	-34.01
Mole number of gases (mol/kg)	24.51	14.48	47.26	31.66	28.96	49.79
Volume of gases (Nm^3/kg)	0.549	0.324	1.059	0.710	0.649	1.116
Composition (mol%)						
CO_2	14.73	4.01	2.45	16.39	10.43	9.84
H_2O	5.06	0.13	87.19	5.58	7.08	68.12
N_2	69.00	15.47	-	76.18	54.05	0.71
CO	-	0.11	0.002	-	0.20	0.12
H_2	-	0.05	0.82	-	1.84	11.19
CH_4	-	-	1.55	-	0.83	7.16
HCl	4.62	33.53	5.99	1.14	2.65	1.43
C (solid)	-	29.77	-	-	20.29	-
As	-	11.18	2.00	-	2.65	1.43
$AsCl_3$	5.82	5.77	0.005	0.08	-	-
As_2O_3	0.77	-	-	0.65	-	-

TABLE 3. S-Lost and N-Lost

Substance	S-LOST		N-LOST	
	with air (1 bar)		with air (1 bar)	
O_2-balance	0.	-30.	0.	-30.
Mole number gases (mol/kg)	32.87	31.11	34.19	34.41
Volume of gases (Nm^3/kg)	0.737	0.697	0.766	0.771
Composition (mol%)				
CO_2	14.40	6.66	12.82	6.99
H_2O	7.23	8.02	11.75	11.89
N_2	67.55	40.31	71.15	50.88
CO	0.002	0.17	-	0.18
H_2	0.01	2.66	-	3.97
CH_4	-	1.67	-	3.50
HCl	7.21	14.01	4.27	8.31
H_2S	3.57	6.99	-	-
COS	0.02	0.02	-	-
C (solid)	-	19.51	-	14.28

TABLE 4. 2-Chloroacetophenone and Bromoacetone

Substance	2-Chloroaceto-phenone		Bromoacetone	
	with air (1 bar)		with air (1 bar)	
O_2-balance	0.	-30.	0.	-30.
Mole number gases (mol/kg)	32.96	31.23	31.01	29.95
Volume of gases (Nm^3/kg)	0.739	0.700	0.695	0.671
Composition (mol%)				
CO_2	17.49	11.63	15.69	9.68
H_2O	6.56	8.38	10.46	12.70
N_2	73.76	51.86	68.62	39.90
CO	-	0.22	-	0.20
H_2	-	2.09	-	3.48
CH_4	-	1.04	-	2.88
HCl	2.19	4.19	-	-
HBr	-	-	5.23	10.98
C (solid)	-	20.60	-	20.17

TABLE 5: Chloropicrin and Phosgene

Substance	Chloropicrin	Phosgene
O_2-balance	14.60	0.
Mole number gases (mol/kg)	18.25	14.613
Volume of gases (Nm^3/kg)	0.409	0.328
Composition (mol%): CO_2	33.33	34.40
N_2	16.67	-
CO	-	0.39
Cl_2	50.00	30.82
CCl_4	-	19.18
C (solid)	-	15.22

TABLE 6. Tabun, Soman and Sarin

Substance	TABUN		SOMAN		SARIN	
	with air (1 bar)		with air (1 bar)		with air (1 bar)	
O_2-balance	0.	-30.	0.	-30.	0.	-30.
Mole number of gases (mol/kg)	32.98	32.58	33.59	33.76	32.86	32.51
Volume of gases (Nm^3/kg)	0.739	0.730	0.753	0.757	0.736	0.729
Composition(mol%)						
CO_2	11.91	7.83	12.23	8.02	11.64	7.81
H_2O	13.10	13.68	13.10	13.62	13.09	13.67
N_2	73.80	55.64	72.04	52.70	70.90	49.78
CO	-	0.19	-	0.19	-	0.19
H_2	-	4.35	-	4.32		4.36
CH_4	-	4.13	-	4.00	-	4.14
HF	-	-	1.75	3.46	2.91	5.85
NH_3	-	0.01	-	0.01	-	0.01
C (solid)	-	10.69	-	11.99	-	11.26
P_2O_5	1.19	-	0.87	-	1.46	-
P_2O_3	-	2.39	-	1.73	-	2.93

TABLE 7. 2,4,6-Trinitrotoluene (TNT)

Substance	TNT			
	pure	with air (1 bar)	with air (1 bar)	80% H_2O (220 bar)
O_2-balance	-73.96	-30.	0.	-14.79
Mole number gases (mol/kg)	25.98	29.99	32.42	49.66
Volume of gases (Nm^3/kg)	0.582	0.672	0.727	1.113
Composition (mol%)				
CO_2	20.18	21.72	22.82	7.97
H_2O	16.50	11.59	8.15	84.12
N_2	14.27	41.95	69.03	2.64
CO	0.24	0.30	-	0.01
H_2	2.63	2.10	-	0.79
CH_4	2.33	1.07	-	4.44
NH_3	-	-	-	0.04
C (solid)	43.86	19.51		

Fig. 1: Diphenylchloroarsine (DA) $C_{12}H_{10}AsCl$
Clark I

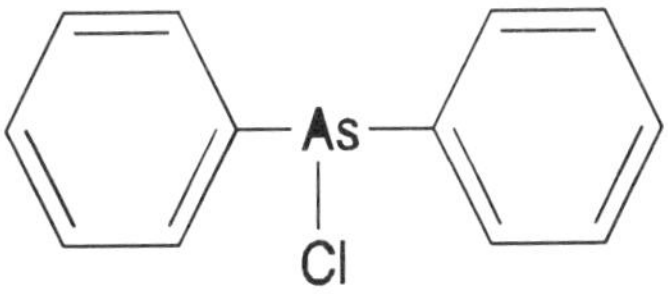

Melting Point [°C]: 44
Boiling Point [°C]: 333
Density [g/cm^3]: 1.422 (20°C)
Volatility [mg/m^3] : 6.8

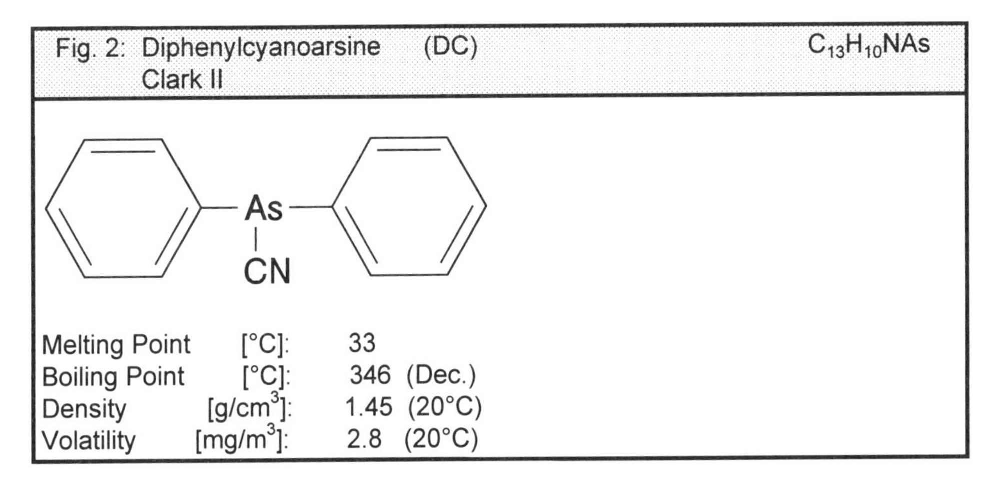

Fig. 2: Diphenylcyanoarsine (DC) $C_{13}H_{10}NAs$
Clark II

Melting Point [°C]: 33
Boiling Point [°C]: 346 (Dec.)
Density [g/cm^3]: 1.45 (20°C)
Volatility [mg/m^3]: 2.8 (20°C)

Fig. 3: 2-Chlorovinyldichloroarsine (Agent L) $C_2H_2AsCl_3$
Lewisite

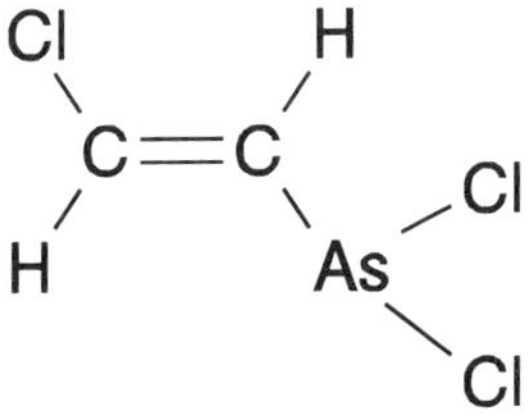

Melting Point [°C]: 0
Boiling Point [°C]: 190
Density [g/cm^3]: 1.888
Volatility [mg/m^3] : 4480 (20°C)

Fig. 4: Diphenylaminechloroarsine (DM) $C_{12}H_9NAsCl$
Adamsite

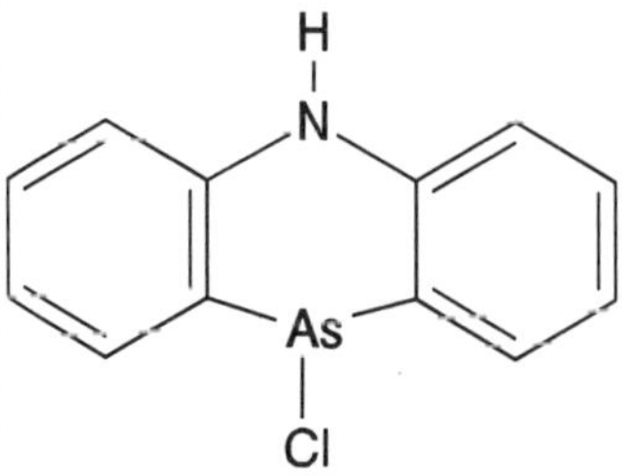

Melting Point	[°C]:	195
Boiling Point	[°C]:	410 (Dec.)
Density	[g/cm^3]:	1.648 (20°C)
Volatility	[mg/m^3]:	0.02 (20°C)

Fig. 5: Bis(2-chloroethyl)-sulfide (HD) $C_4H_8Cl_2S$
Mustard gas, Lost

$$Cl-CH_2-CH_2-S-CH_2-CH_2-Cl$$

Melting Point	[°C]:	14.4
Boiling Point	[°C]:	217.5 (Dec.)
Density	[g/cm^3]:	1.274 (20°C)
Volatility	[mg/m^3]:	628 (20°C)

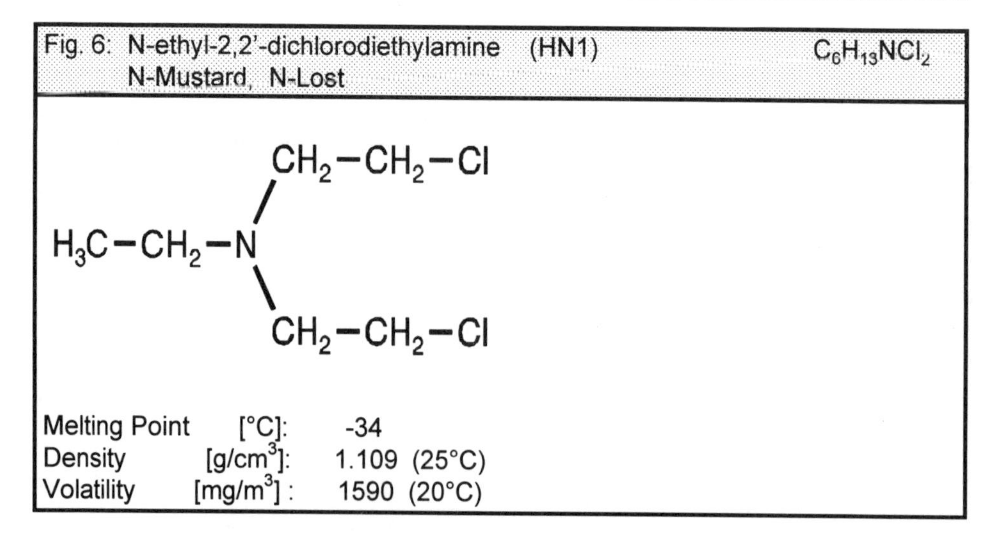

Fig. 6: N-ethyl-2,2'-dichlorodiethylamine (HN1) $C_6H_{13}NCl_2$
N-Mustard, N-Lost

Melting Point	[°C]:	-34
Density	[g/cm^3]:	1.109 (25°C)
Volatility	[mg/m^3]:	1590 (20°C)

Fig.7: 2-Chloroacetophenone (CN) C_8H_7OCl

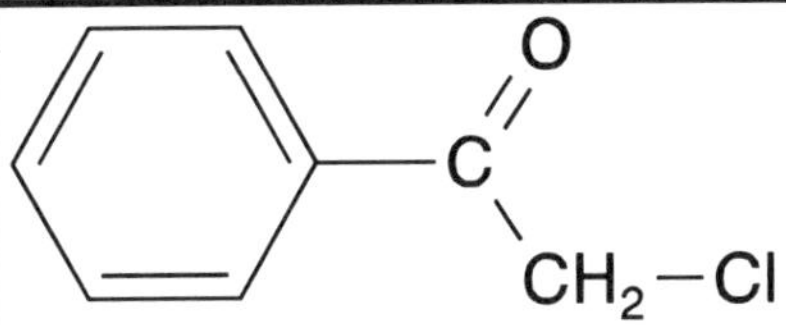

O
C
CH_2 — Cl

Melting Point [°C]: 54 - 59
Boiling Point [°C]: 245
Density [g/cm^3]: 1.321
Volatility [mg/m^3] : 105 (20°C)

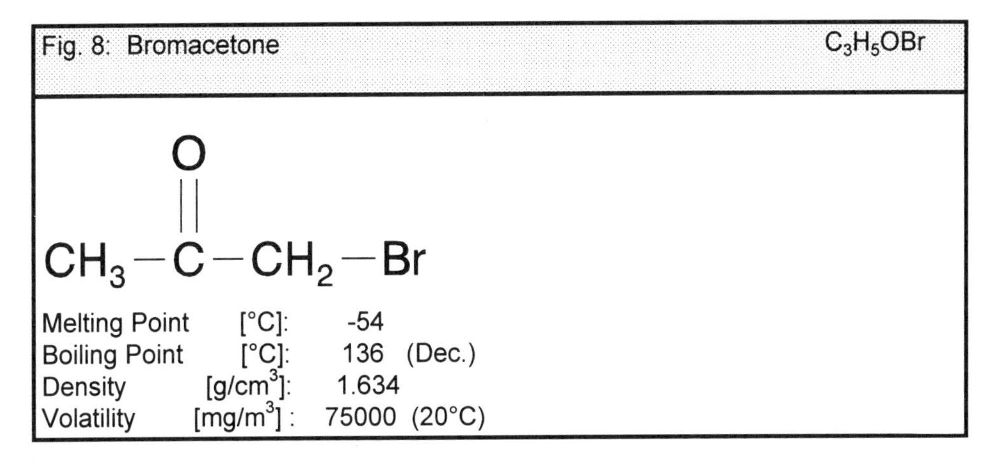

Fig. 8: Bromacetone C_3H_5OBr

Melting Point [°C]: -54
Boiling Point [°C]: 136 (Dec.)
Density [g/cm^3]: 1.634
Volatility [mg/m^3] : 75000 (20°C)

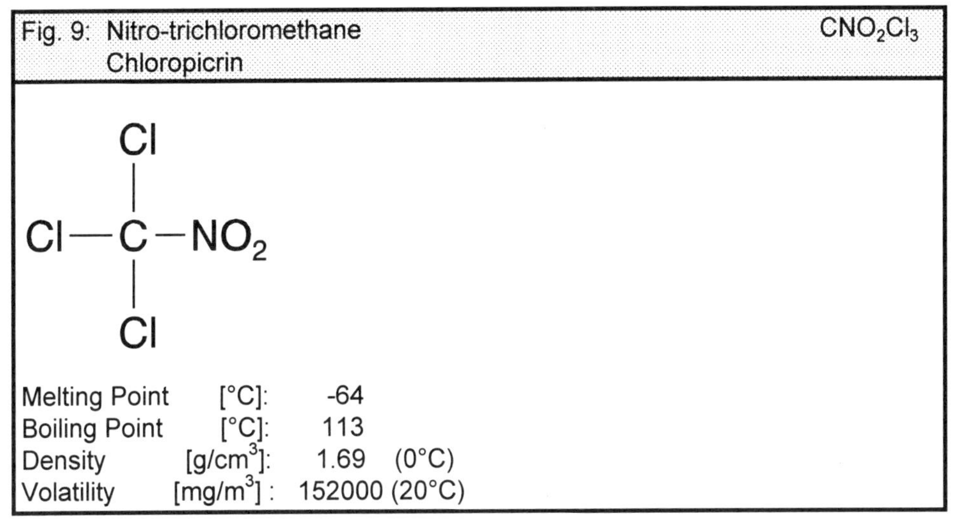

Fig. 9: Nitro-trichloromethane CNO_2Cl_3
Chloropicrin

Melting Point [°C]: -64
Boiling Point [°C]: 113
Density [g/cm^3]: 1.69 (0°C)
Volatility [mg/m^3] : 152000 (20°C)

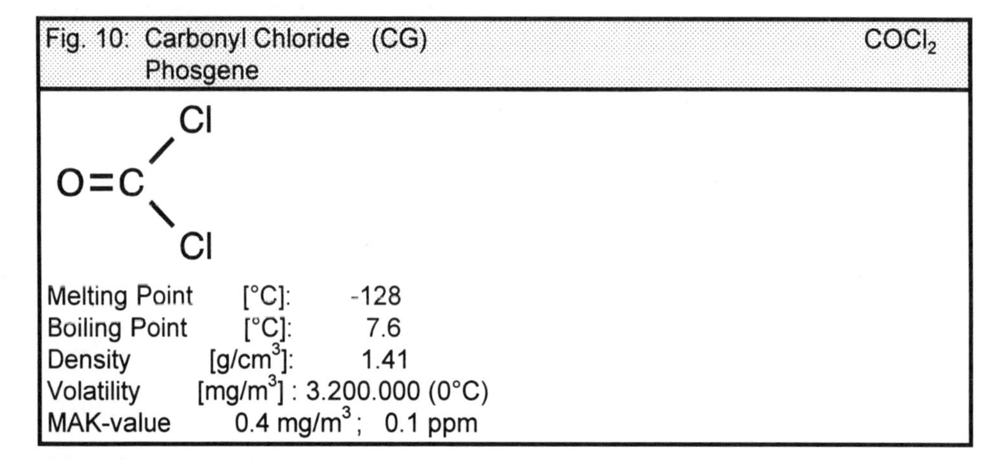

Fig. 10: Carbonyl Chloride (CG) Phosgene — $COCl_2$

Melting Point [°C]: -128
Boiling Point [°C]: 7.6
Density [g/cm^3]: 1.41
Volatility [mg/m^3] : 3.200.000 (0°C)
MAK-value 0.4 mg/m^3; 0.1 ppm

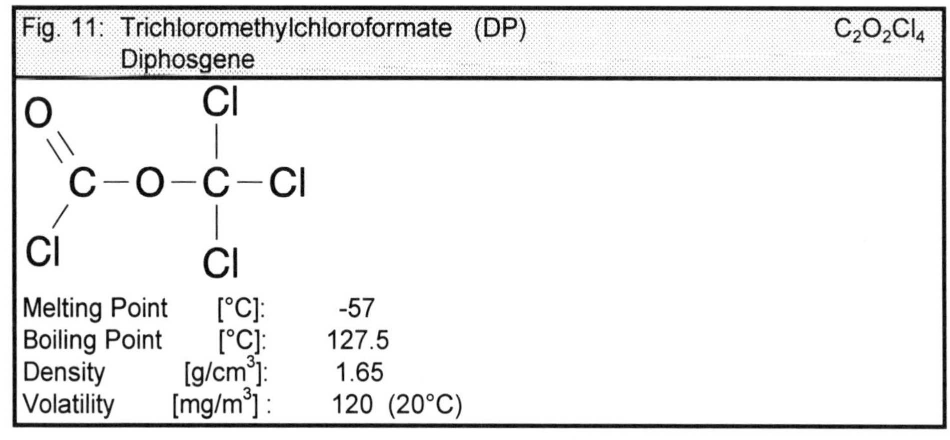

Fig. 11: Trichloromethylchloroformate (DP) Diphosgene — $C_2O_2Cl_4$

Melting Point [°C]: -57
Boiling Point [°C]: 127.5
Density [g/cm^3]: 1.65
Volatility [mg/m^3] : 120 (20°C)

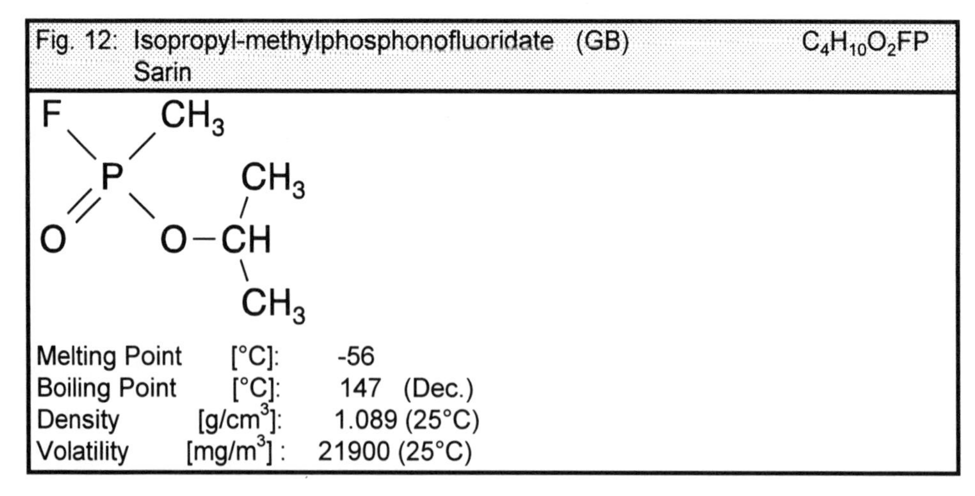

Fig. 12: Isopropyl-methylphosphonofluoridate (GB) Sarin — $C_4H_{10}O_2FP$

Melting Point [°C]: -56
Boiling Point [°C]: 147 (Dec.)
Density [g/cm^3]: 1.089 (25°C)
Volatility [mg/m^3] : 21900 (25°C)

Fig. 13: Pinacolyl-methylphosphonofluoridate (GD) Soman — C7H16O2FP

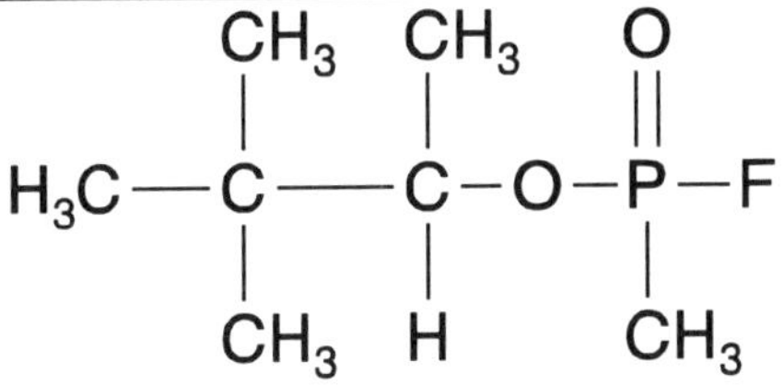

Melting Point [°C]:	-80
Boiling Point [°C]:	167
Density [g/cm^3]:	1.022 (25°C)
Volatility [mg/m^3] :	3060 (25°C)

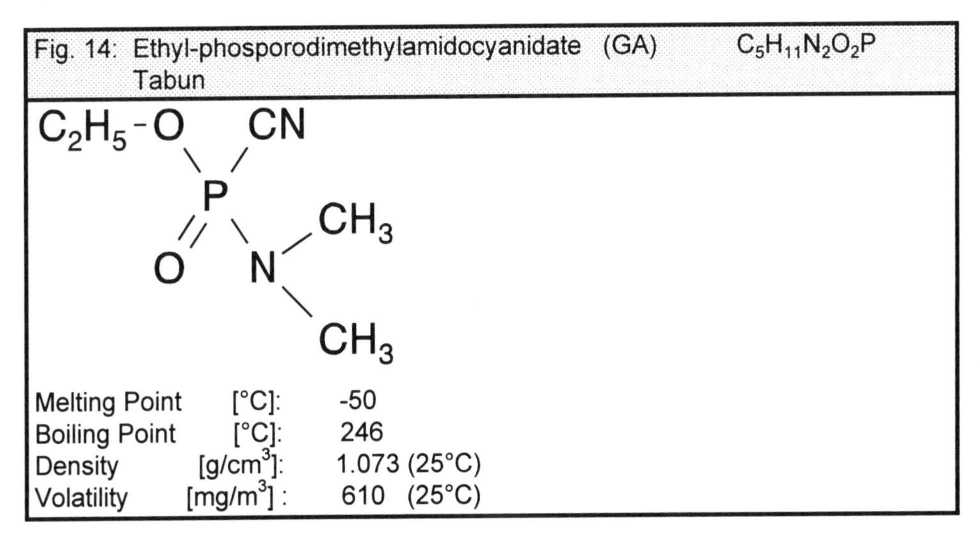

Fig. 14: Ethyl-phosporodimethylamidocyanidate (GA) Tabun — $C_5H_{11}N_2O_2P$

Melting Point [°C]:	-50
Boiling Point [°C]:	246
Density [g/cm^3]:	1.073 (25°C)
Volatility [mg/m^3] :	610 (25°C)

Fig. 15: 2,4,6 - Trinitrotoluene (TNT) — $C_7H_5N_3O_6$

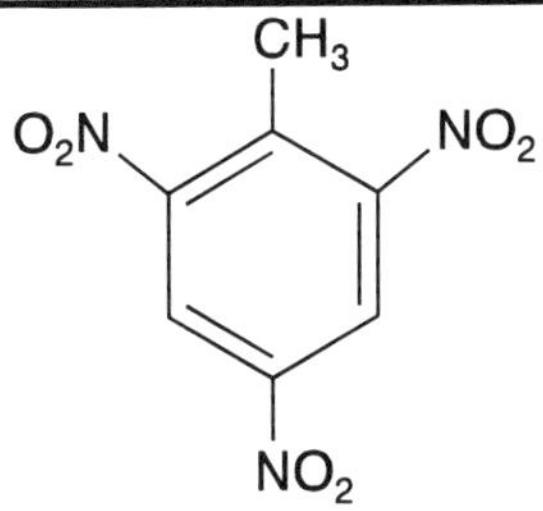

Melting Point [°C]:	80.7
Density [g/cm^3]:	1.654
Oxygen balance [%] :	-73.96

BACTERIAL BIODEGRADATION OF NITRATE ESTER EXPLOSIVES

GRAHAM F. WHITE, JASON R. SNAPE
Department of Biochemistry
University of Wales College of Cardiff
PO Box 903, Cardiff CF1 1ST, United Kingdom
S. NICKLIN
Defence Research and Evaluation Agency
Fort Halstead, Sevenoaks, Kent, TN14 7BP, UK

Abstract

Besides toxic agents, chemical weapons munitions also contain propellants and explosives. Moreover, sea-dumping also included conventional munitions containing high explosives. Nitro-compounds such as TNT, and nitrate esters such as glycerol trinitrate (GTN) and pentaerythritol tetranitrate (PETN) which have been widely used in munitions, are potential sources of nutrition for bacteria. This paper describes recent work on the bacterial biodegradation of GTN and PETN by bacteria isolated from rivers, soil and sewage. The capacity of bacteria to biodegrade explosives has implications for the fate of explosives released from corroded munitions, for the development of remediation/disposal strategies for recovered munitions, and in the development of sensitive biosensor detectors for surveying and monitoring dump-sites.

1. Introduction

Nitrate esters of aliphatic alcohols are industrially important as a result of two major, though incongruous, applications. First they are widely used as powerful explosives, and secondly they are effective vasodilators in the treatment of heart disease such as *angina pectoris*. Glycerol trinitrate (GTN; propane-1,2,3,-triyl trinitrate) and pentaerythritol tetranitrate (PETN; 2,2-bis[nitrooxymethyl]propane-1,3-diyl dinitrate) are among the most widely known and are used in both applications. They are produced by direct nitration of the parent alcohol with nitric acid, usually with sulphuric acid present as catalyst (Urbanski, 1965 [1]), and may be regarded as the nitric acid tri- and tetra-esters, respectively, of glycerol and pentaerythritol (Fig. 1). Thus structurally these nitrate ester groups are analogous to phosphate and sulphate esters. However it has recently been emphasised that whereas the latter groups occur ubiquitously in the biosphere, nitrate esters have never been detected as naturally-occurring compounds in

A. V. Kaffka (ed.), Sea-Dumped Chemical Weapons: Aspects, Problems and Solutions, 145–156.

living organisms and thus constitute a true xenobiotic challenge to biological systems (White & Snape, 1993 [2]). The microbial biodegradability of nitrate esters is therefore not only academically intriguing but is also relevant to development of methods for the elimination of nitrate esters from production waste-waters and for the bioremediation of land-sites contaminated during production, storage and use of such compounds (Kaplan, 1992 [3]; Walker & Kaplan, 1992 [4]; Gorontzy *et al.*, 1994 [5]).

Before embarking on a search for bacteria competent in the biodegradation of nitrate esters, it is helpful first to appreciate *why* they might do so. Low-nutrient (oligotrophic) environments are often considered among the "extremes" to which micro-organisms have become adapted. Yet, oligotrophic conditions are so ubiquitous in the microbial biosphere, as to be the norm (Morgan and Dow, 1986 [6]). Moreover, nutrient status is determined not only by carbon availability but also by other bio-elements (predominantly P, S, N), and in oligotrophic environments it is likely that microbial growth is constrained by all these, either simultaneously, or sequentially. Faced with such privation, micro-organisms have evolved numerous strategies for surviving nutrient limitation, including chemotaxis, surface-attachment, and in more recent times diversification of metabolic capabilities to include catabolism of the industrial products of human-kind.

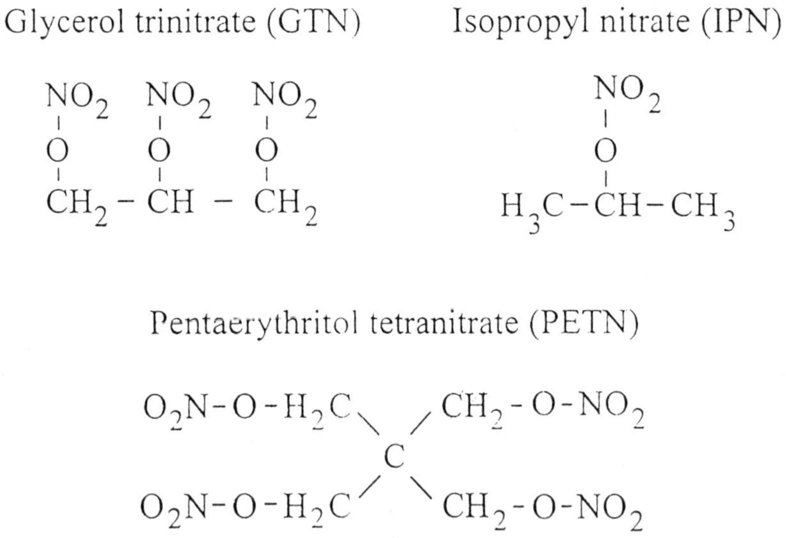

Glycerol trinitrate (GTN)

$$\begin{array}{ccccc} NO_2 & & NO_2 & & NO_2 \\ | & & | & & | \\ O & & O & & O \\ | & & | & & | \\ CH_2 & - & CH & - & CH_2 \end{array}$$

Isopropyl nitrate (IPN)

$$\begin{array}{c} NO_2 \\ | \\ O \\ | \\ H_3C-CH-CH_3 \end{array}$$

Pentaerythritol tetranitrate (PETN)

$$\begin{array}{ccc} O_2N-O-H_2C & & CH_2-O-NO_2 \\ & \diagdown \quad \diagup & \\ & C & \\ & \diagup \quad \diagdown & \\ O_2N-O-H_2C & & CH_2-O-NO_2 \end{array}$$

Fig. 1. Structures of GTN, PETN and IPN.

Examination of the structures of nitrate ester explosives shows that they contain abundant carbon derived from polyhydric alcohols (e.g. glycerol, pentaerythritol) which are readily assimilable by bacteria. Moreover nitrate esters also contain abundant nitrogen and thus they may also serve to supply this element for microbial growth. A very few bacterial species can fix dinitrogen from the air to synthesise amino acids and other organonitrogen compounds, but many more can reduce nitrate and/or nitrite to ammonia prior to its assimilation into amino acids. A further group (the denitrifiers) use nitrate or nitrite as an oxidising agent in place of dioxygen to oxidise organic fuels under anaerobic conditions to yield energy (Payne, 1981 [7]). If there are routes for utilisation of nitrate ester nitrogen, these involving nitrate and/or nitrite seem the most likely.

The well-documented metabolism of GTN in mammalian systems (Taylor *et al.*, 1987 [8]) involves conversion to nitric oxide (Feelisch & Noack, 1987 [9]; Schror *et al.*, 1991 [10]) which currently enjoys new-found importance in mammalian physiology as a chemical messenger with roles in vascular function (Moncada *et al.*, 1988 [11]), neurotransmission (Knowles *et al.*, 1989 [12]) and immune response (Knowles *et al.*, 1990 [13]). Fungi such as *Phanerochaete chrysosporium* and *Geotrichum candidum* also metabolise GTN and are serving as model systems for the study of eukaryotic metabolism of nitrate esters (Ducrocq *et al.*, 1989, 1990 [14, 15]; Servent *et al.*, 1991 [16]). Fungal fission of the nitrate ester linkage is also reductive with the formation of nitrite and NO-protein complexes. Published work on bacterial metabolism of GTN is confined to an earlier study (Wendt *et al.*, 1978 [17]) which established sequential denitration via glycerol dinitrates (GDN) and glycerol mononitrates (GMN) to glycerol.

The present work was undertaken as part of a wider programme to isolate bacteria competent in the biodegradation of nitrate esters and to determine the kinetics of metabolite production during the biodegradation process. The enrichment substrates used were GTN, PETN and propane-2-yl nitrate (isopropyl nitrate, IPN). The last was chosen because it is a readily available mononitrate which is structurally related to GTN, and it is analogous to propane-2-yl sulphate which is known to undergo bacterial biodegradation (Crescenzi *et al.*, 1984, 1985 [18, 19]).

2. Materials and Methods

2.1. CHEMICALS

Three esters, *viz.* GTN, PETN and isopropyl nitrate (IPN), were used for the initial isolation of bacteria. Research quantities of GTN (adsorbed on Kieselguhr, 20% w/w), glycerol 1-mononitrate (1-GMN), glycerol 1,2-dinitrate (1,2-GDN), glycerol 1,3-dinitrate (1,3-GDN) and PETN were supplied by Fort Halstead, Sevenoaks, Kent. IPN was purchased from Aldrich Chemicals, Poole, Dorset.

Working solutions of GTN were prepared by extracting 20 mg of GTN-on-Kieselguhr into 0.5 ml ethanol. The mixture was shaken vigorously for 15 min, then centrifuged at 2500 rpm for 5 min. The supernatant (0.4 ml) was used for the preparation of microbial growth media.

HPLC eluents were prepared using HiperSolv quality methanol (Merck-BDH) and ultra-pure water from a Millipore Milli-Q 50 system which produced 0.22 μm-filtered, 18 Mohm deionised water.

Other chemicals were the purest available from Merck (BDH) or Sigma Chemical Co.

2.2. ISOLATION AND CULTURING OF COMPETENT BACTERIA

Standard enrichment culturing techniques were used to isolate bacteria from a variety of sources (see Table 1) for ability to use one of the three nitrate esters GTN, PETN, and IPN as source of nitrogen for growth. Samples of inoculum material were added to culture flasks containing a minimal salts medium lacking nitrogen (see below) and 1% v/v glycerol as carbon source, plus the appropriate nitrate ester as sole source of nitrogen. Flasks were incubated in a gyratory incubator at 25oC for 1-2 weeks and samples were sub-cultured at weekly intervals thereafter into fresh medium. After 4-8 sub-cultures, samples from flasks showing visible growth were plated out onto nutrient agar plates and single colony-types separated and re-plated to purity. Axenic strains were re-inoculated into the basal salts/glycerol/nitrate ester medium to check that they had retained the capacity for growth.

TABLE 1. Number of isolates obtained from various sources by enrichment on different nitrate esters

Source of organisms	Nitrate ester		
	GTN	IPN	PETN
Sewage (SW)	2	0	4
Soil (SL)	3	0	4
River (RI)	1	0	4
Rat faeces (RF)	0	0	2

For isolation and subsequent culturing, the basal salts medium contained (per litre) K_2HPO_4, 3.5 g; KH_2PO_4, 1.5 g; NaCl, 0,5 g; $MgSO_4$, 0.12; 1 ml of trace elements solution. Trace elements solution contained (per litre) sodium borate, 0.57; $FeCl_3.6H_2O$, 0.24g; $CoCl_2.6H_2O$, 0.04g; $CuSO_4.5H_2O$, 0.06g; $MnCl_2.4H_2O$, 0.03g; $ZnSO_4.7H_2O$, 0.31g; $Na_2MoO_4.2H_2O$, 0.03g; this solution was stored in the dark at 4°C. Culture flasks were prepared by adding 2 ml glycerol to 100 ml basal salts medium immediately before autoclaving. After autoclaving and cooling, 0.4 ml of ethanolic GTN working solution was added to give the desired final concentration of GTN as sole added source of nitrogen for growth. Cultures were agitated at 100 rpm and 25°C and growth of bacteria was monitored by measuring optical density at 450 nm in Pharmacia/LKB Novaspek II spectrophotometer.

2.3. STORAGE AND MAINTENANCE OF MICRO-ORGANISMS

Bacterial isolates were maintained on working and short-term back-up nutrient agar slopes stored at 4°C. Cultures were plated out every 2 months to ensure purity, and fresh slopes made. Long-term storage was at -70°C in glass embroidery beads with 15% glycerol in nutrient broth as cryo-protectant (Jones *et al.*, 1991 [20]).

Geotrichum candidum was obtained from Birkbeck College Culture Collection and maintained on nutrient agar plates with 2-monthly transfers.

2.4. ANALYSIS OF NITRATE ESTERS AND THEIR METABOLITES

Residual GTN and the di- and mono-nitrated analogues arising from bacterial biodegradation were analysed by HPLC using a Dionex DX300 Series Ion Chromatograph (Dionex UK, Camberley) equipped with a variable wavelength UV detector and a Spectra Physics 4400 Integrator. The column was a Lichrosorb ODS 10 µm-bead size, 260 x 4.6 mm (Phase Separations, Clwydd, Deeside) eluted with a water/methanol linear gradient from 5% to 50% methanol over 30 minutes. Eluents were prepared from HPLC grade solvents and degassed by sparging with helium and maintaining under a helium atmosphere. Analytes were detected by their absorbency at 217 nm.

HPLC retention times for GTN and glycerol 1-mononitrate (1-GMN) were determined by elution of authentic materials, and calibration curves were obtained using standard solutions, the concentrations in which were determined using published extinction coefficients (Dunstan *et al.*, 1965 [21]). Calibration of the HPLC system for glycerol dinitrates presented some problems because the chemical preparation of pure samples is difficult. Thus the available sample of "1,2-GDN" contained 2 peaks and it was impossible to assign one or other to 1,2-GDN. To circumvent this difficulty, a mixture of the dinitrates was prepared from standard solutions of GTN by incubation with *Geotrichum candidum.* This micro-organism is known (Ducrocq *et al.*, 1989 [14]) to yield a mixture of dinitrates but with the 1,3,-isomer dominating over the 1,2-isomer, thus enabling the HPLC peaks to be identified by relative size. Calibration was achieved from the known initial and measured (HPLC) residual concentrations of GTN. The available sample of 1,3-GDN contained 4 peaks, 3 of which were identified with 1-GMN, 1,2-GDN and 1,3-GDN. The fourth peak, eluting between 1-GMN and the dinitrates, was tentatively attributed to 2-GMN.

Nitrite was detected using the Greiss-Romijin reagent (sulphanilic acid/ N-1-naphthylethylenediamine in acid) as described by Tan-Walker (Tan-Walker, 1987 [22]).

3. Results

3.1. ISOLATION OF BACTERIA

Table 1 summarises the distribution of isolates obtained from each of the source samples, using each nitrate ester separately as source of nitrogen. Of the three esters used, GTN and PETN appeared to be the most readily biodegradable, judging by the number of isolates obtained. No isolates were obtained which could utilise IPN. Of the various source samples, the sewage, soil and river samples were the most productive.

Some of the isolates in Table 1 were selected for further study. The remainder of this paper is devoted to the results obtained for strain RI-NG1, isolated from river sediment for its ability to biodegrade GTN, and tentatively identified as a *Pseudomonas* sp. using the BIOLOG bacterial identification (BIOLOG, Hayward, Ca, USA) system.

3.2. DEPENDENCE OF GROWTH ON ESTER AVAILABILITY

In order to establish that disappearance of nitrate esters was the result of microbial metabolism, experiments were initiated to show that growth was dependent on the presence of nitrate esters. Cells pre-adapted by growth on basal salts, glycerol/ GTN (250μM)) medium for 72h, were used to inoculate (0.1% v/v) a second flask, designated A and containing 200 ml of the same medium. Growth in flask A was monitored for 240h. *Pseudomonas* sp. RI-NG1 grew with a doubling time of about 24h, reaching a final OD_{450} of about 0.4 (Fig. 2). In view of the high concentration of glycerol and low concentration of GTN, it was likely that growth was nitrogen-limited. To confirm this, at 120h the culture in flask A was sub-divided as follows:- 100 ml of the residual culture was transferred aseptically to a clean sterile flask (B) and supplemented with a further aliquot of GTN; 1 ml aliquots from flask A were used to inoculate flask C which contained fresh sterile whole media (i.e. including GTN), and flask D which contained fresh sterile media lacking GTN; the residual culture medium in flask A was left untreated. All four flasks A-D were returned to the incubator and monitored for growth over a further 5-day period. Fig. 2 shows that after the initial 120h period, the culture in flask A was entering stationary phase. Addition of fresh GTN to the culture at this point (flask B) led to a second burst of growth which approximately doubled the cell density. Dilution (100-fold) of the first culture into fresh whole medium (flask C) gave a very similar burst of growth, suggesting that growth in the flask A at 120h was limited only by GTN availability. Transfer from flask A to flask D lacking GTN produced no growth at all.

3.3. METABOLITE PRODUCTION

The samples removed from flask A for measurements of growth, were also analysed by HPLC for GTN and its metabolites (Fig. 3). Growth was accompanied by disappearance of GTN which was complete at 96h, and by appearance of both isomers

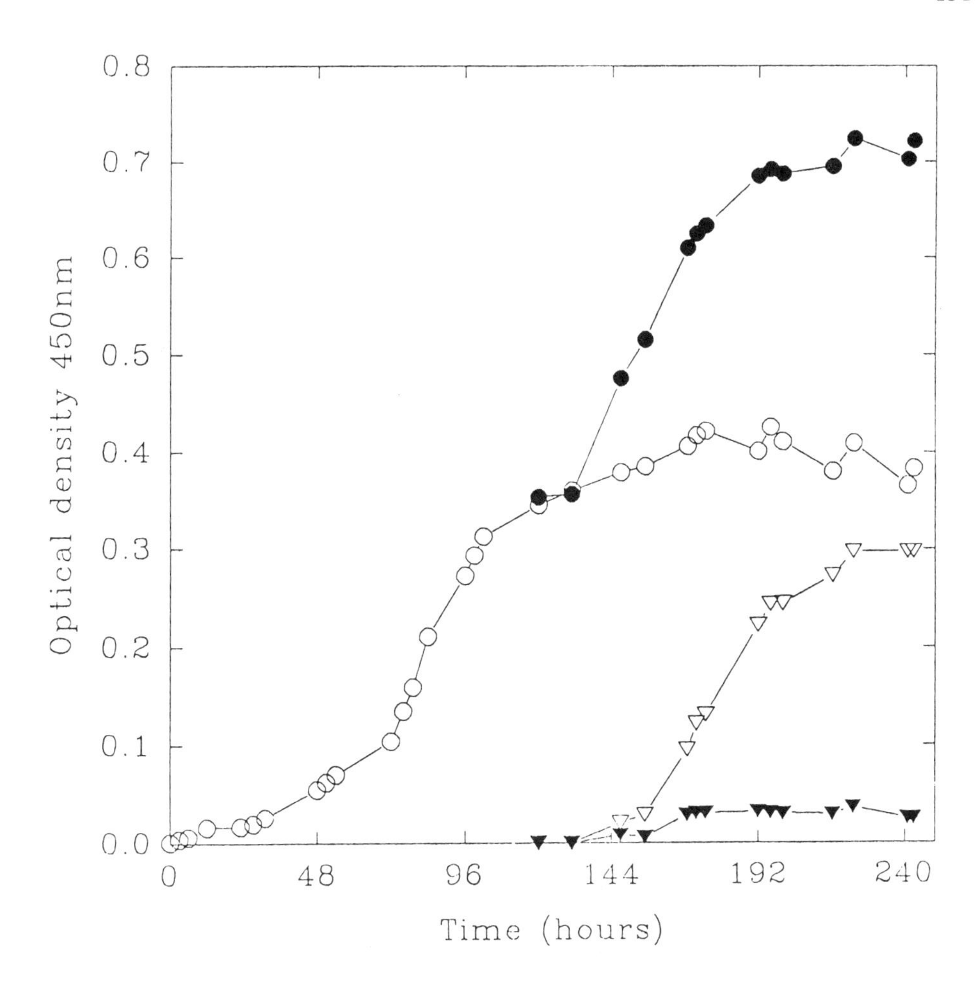

Fig. 2. Growth of *Pseudomonas* sp. RI-NG1 in basal salt, glycerol medium with GTN as sole added source of nitrogen. Flask A (O) containing 200 ml basal salts/ 1% glycerol and 250 μM GTN, was inoculated from a culture of *Pseudomonas* sp. RI-NG1 in similar medium. After 120 h, 100 ml was transferred from flask A to a clean flask B (●) and supplemented with a further aliquot of GTN. At the same time, flask C (▽) containing fresh whole medium and flask D (▼) containing fresh medium lacking GTN, were inoculated from flask A (0.1% v/v). Growth was monitored by measuring OD_{450}.

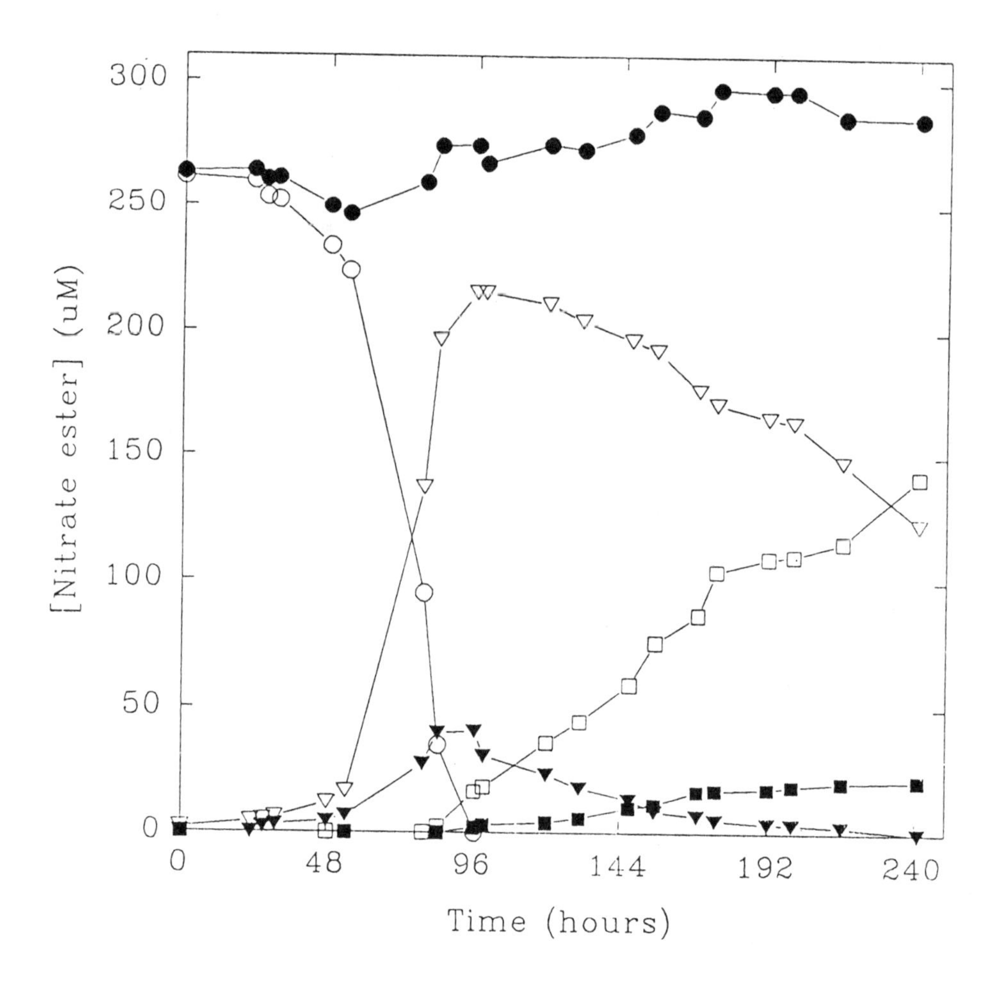

Fig. 3. Production of metabolites from GTN by *Pseudomonas* sp. RI-NG1: O, GTN; ∇, 1,3-GDN; ▼, 1,2-GDN; □, 1-GMN; ■, 2-GMN; ●, total residual GTN and its metabolites.

of GDN. 1,3-GDN was the major metabolite, its rate of appearance exceeding that of the 1,2 isomer by about 5-fold. 1-GMN and 2-GMN first appeared after about 80h, and increased in concentration at the expense of decreasing concentrations of the dinitrates throughout the remainder of the experiment. The rate of production of 1-GMN exceeded that of the 2-isomer by about 6-fold. Fig. 3 also shows that the total concentration of all the glycerol nitrates detected was constant throughout the 240h period of incubation.

Preliminary experiments using broth-grown resting cells incubated with GTN showed that disappearance of GTN was accompanied by formation of nitrite ions in solution.

4. Discussion

The growth experiments demonstrated unequivocally that GTN was serving as the sole source of nitrogen for growth of *Pseudomonas* sp. RI-NG1 under these conditions. During growth, GTN was converted stoichiometrically (Fig. 3) to its di-nitrate analogues which were produced simultaneously but with 1,3-GDN appearing about 5 times faster than the 1,2-isomer. This is significantly higher than the ratio of 1.2-1.8 for 1,3-GDN/1,2-GDN produced by *G. candidum* (Ducrocq *et al.*, 1989 [14]). Moreover, bearing in mind that GTN offers only one way of making 1,3-GDN (denitration at the central nitrate) but two ways for making 1,2-GDN (attack at either end) the selectivity for attack at the central nitrate group (C2) is about ten-times that at the terminal esters.

When all the GTN had disappeared, and the GDNs were approaching their maximum concentrations, the mono-nitrates 1-GMN and 2-GMN began to appear, again simultaneously but in the ratio of about 6:1. Whichever of its nitrate esters is removed, 1,3-GDN can only give rise to 1-GMN. Thus because 1,3-GDN is the dominant intermediate, the bulk of the 1-GMN production (Fig. 3) is attributable to the biodegradation of 1,3-GDN. In contrast, denitration of 1,2-GDN can produce either 1-GMN or 2-GMN depending on whether the attack occurs at C2 or C1 respectively. After 96 h, GTN degradation was complete and no more dinitrates could be made. Examining the data after this point in Fig. 3, it is clear that most but not all of the 1,2-GDN is converted to 2-GMN, indicating a preference for attack at C1 in 1,2-GDN. Thus the regioselectivity for C1/C2 denitration in GTN is different from that in 1,2-GDN. Whether this difference reflects either the operation of different enzyme systems or different modes of substrate binding to a common enzyme, remains to be determined.

The total recovery of residual GTN plus all isomers of GDN and GMN was close to 100% throughout the period of the experiment. This indicated that no other GTN derivative or metabolite was produced in significant quantities and also suggested that the organism was incapable of biodegrading either of the mononitrates.

Based on the data presented herein, Fig. 4 summarises the sequence of metabolite production and indicates the relative contributions of individual steps to the overall

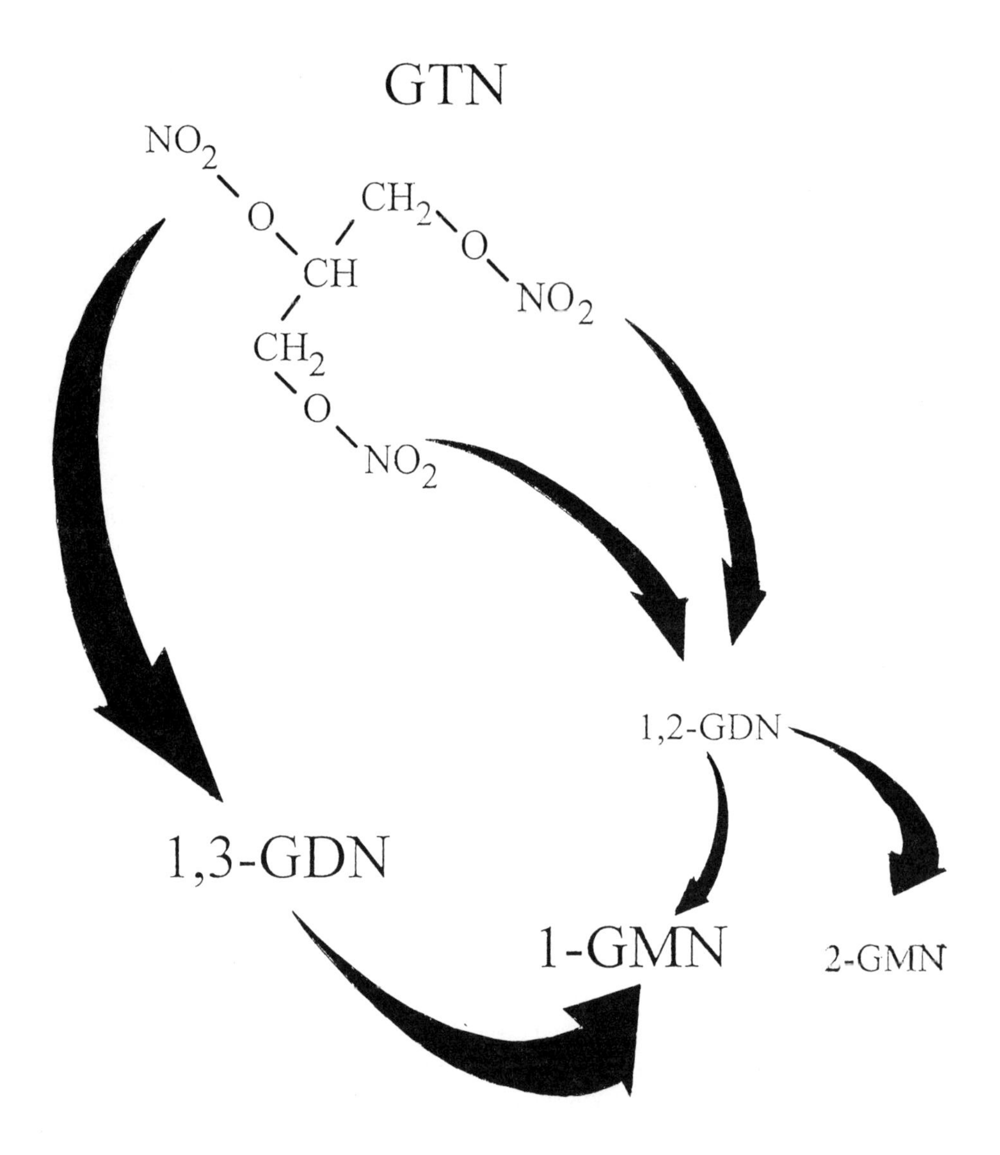

Fig. 4. Pathway for the biodegradation of GTN in *Pseudomonas* sp. RI-NG1. The thickness of each arrow gives an approximate guide to the contribution of that step to the overall pathway.

biodegradation pathway. The detection of nitrite in incubations of resting cells with GTN is consistent with a reductive mechanism enabling the ester nitrogen to enter the assimilation pathway at the level of nitrite. Although not yet tested, denitrifiers may also be able to denitrify nitrate esters via nitrite to dinitrogen gas.

References

1. Urbanski, T. (1965). *Chemistry and Technology of Explosives,* PWN-Polish Scientific Publishers and Pergamon Press, Warsaw and Oxford.
2. White, G.F. & Snape, J.R. (1993). *J. Gen. Microbiol.*, **139**, 1947-1957.
3. Kaplan, D.L. (1992). *Curr. Opin. Biotechnol.*, **3**, 253-260.
4. Walker, J.E. & Kaplan, D.L. (1992) *Biodegradation* **3**, 369-385.
5. Gorontzy, T., Drzyzga, O., Kahl, M.W., Bruns-Nagel, D, Breitung, J., von Loew, E. & Blotevogel, K.-H. (1994). *Crit. Rev. Microbiol.,* **20**, 265-284.
6. Morgan, P. & Dow, C.S. Bacterial adaptations for growth in low nutrient environments. *In*: Microbes In Extreme Environments, (Herbert, R.A. & Codd, G.A., Eds), 1986, pp.187-214, Academic Press, London.
7. Payne, W.J. (1981) *Denitrification.* New York. Wiley.
8. Taylor, T., Taylor, I.W., Chasseaud, L.F. & Bonn, R. (1987). In *Progress in Drug Metabolism*, ed. J. W. Bridges, L. F. Chasseaud and G. G. Gibson, Taylor and Francis, London, **10**, 207-386.
9. Feelisch, M. & Noack, E.A. (1987). *Biochem. Pharmacol.*, **139**, 19-30.
10. Schror, K., Woditsch, I. & Forster, S. (1991). *Blood Vess.*, **28**, 62-66.
11. Moncada, S., Radomski, M.W. & Palmer, R.M.J. (1988). *Biochem. Pharmacol.*, **37**, 2495-2501.
12. Knowles, R.G., Pallacios, M., Palmer, R.M.J. & Moncada, S. (1989). *Proc. Natl. Acad. Sci. U.S.A.*, **86**, 5159-5162.
13. Knowles, R.G., Salter, M., Brooks, S.L. & Moncada, S. (1990). *Biochem. Biophys. Res. Comm.*, **172**, 1042-1048.
14. Ducrocq, C., Servy, C. & Lenfant, M. (1989). *FEMS Microbiol. Lett.*, **65**, 219-222.
15. Ducrocq, C., Servy, C. & Lenfant, M. (1990). *Biotechnol. Appl. Biochem.*, **12**, 325-330.
16. Servent, D., Ducrocq, C., Henry, Y., Guissani, A. & Lenfant, M. (1991). *Biochim. Biophys. Acta*, **1074**, 320-325.
17. Wendt, T.M., Cornell, J.H. & Kaplan, A.M. (1978). *Appl. Environ. Microbiol.*, **36**, 693-699.
18. Crescenzi, A.M.V., Dodgson, K.S. & White, G.F. (1984). *Biochem. J.*, **223**, 487-494.
19. Crescenzi, A.M.V., Dodgson, K.S., White, G.F. & Payne, W.J. (1985). *J. Gen. Microbiol.*, **131**, 469-477.
20. Jones, D., Pell, P.A. & Sneath, P.H.A. (1991). In *Maintenance of microorganisms and cultured cells*, ed. B. E. Kirsop and A. Doyle, Academic Press, London, pp. 45-50.

21. Dunstan, I., Griffiths, J.V. & Harvey, S.A. (1965). *J. Chem. Soc.*, **120**, 1325-1327.
22. Tan-Walker, R.L.B. (1987). *Techniques for analysis of explosive vapours*, PhD thesis, University of London

CHEMICAL WEAPONS DUMPING AND WHITE SEA CONTAMINATION

SERGEY S. YUFIT
N.D. Zelinsky Institute of Organic Chemistry
Russian Academy of Sciences, Moscow;
Centre for Independent Ecological Programs
Socio-Ecological Union, Moscow, Russia
IGOR V. MISKEVICH
Arkhangelsk Ecological Association "Viking", Arkhangelsk, Russia
OLGA N. SHTEMBERG
Ministry of Nature Protection of Russian Federation, Moscow, Russia

This talk consists of three parts. First, a discussion of available data on chemical weapons (CW) dumping in the White Sea. Second, an attempt to analyse the causes of the ecological catastrophe in the White Sea in the summer of 1990. Third, an assessment of the danger from a seal failure on the sunken CW.

1. The Data on CW Dumping in the White Sea

Before we start discussing the contamination of the White Sea as a result of CW dumping, it is necessary to prove that it, in fact, actually exists.

- Military organisations do not give any data on the acts of CW sinking.

- Evidence from participants in CW dumping:

1954. "[In Severodvinsk] for half a year, shells with war gases were being loaded on barges, then the barges were being submerged in the White Sea". As "barges" are mentioned, it indicates that actual submerging took place not far from Severodvinsk, which is in the White Sea.

1956. "From February till May, chemical ammunition (mines, shells and so on) was being taken from the railway station Obozerskaya to Severodvinsk. At first, chemical ammunitions were being sunk somewhere near Severodvinsk, then they were taken to an area close to Spitzbergen."

In this evidence it is important to note that in February, March and April navigation in the Barents Sea (and especially in the Kara Sea) is impeded by ice and ships require an ice-breaking convoy.

A. V. Kaffka (ed.), Sea-Dumped Chemical Weapons: Aspects, Problems and Solutions, 157–166.

As a rule, ice in the Dvina Gulf retreats in the middle of May [1]. Consequently, part of the CW were submerged in the White Sea.

The data on CW dumping near the Cape of Zhelanie, which is to the north of Novaya Zemlya (port of departure Pechenga), are not denied by the administration.

- Official data:

In an interview given by Representative of President of Russia in the Arkhangelsk Region (02/21/92) it was reported that "dumping of CW in the White Sea was carried out from 1947 to 1955." Analysis of his statement allows us to make several important conclusions:

1. The fact that CW dumping occurred in the White Sea is admitted.
2. The period - 8 years (till 1955) - is indicated. Evidence from participants extends this period to 1956, and the other participant's information (from 1954) is confirmed as it falls into the indicated time period.
3. The co-ordinates of the dumpings are given. They coincide with two points of "explosives dumping" indicated on navigational maps of the White Sea. These points are to the north-east of the Solovetzkie Islands.
4. The composition of the dumped CW is given: "700 air bombs and over 5 tonnes (5,453 kg) of a mustard gas-lewisite mixture in 31 iron barrels [i.e. 176 kg in one barrel]." It is obvious that these very precise figures are ridiculously low for the 8-year period of CW dumping, and they do not coincide completely with the evidence from participants in the 1954 and 1956 dumpings on the multi-month unloading of echelons of CW.

1.1. GENERAL CONCLUSION

CW dumping in the White Sea was being carried out for at least 8-9 years. It is a challenge to estimate the amount of CW dumped. They were comprised of obsolete chemical ammunition, i.e. mostly yperite (mustard gas), lewisite and their mixtures. Dumping points indicated officially may not give a true and complete picture.

2. Ecological Catastrophes in the White Sea

2.1. THE ECOLOGICAL CATASTROPHE

An ecological catastrophe (EC) can be defined as an event leading to a noticeable change in the ecological balance of a region, irreversible for a number of years.

From this point of view, the beaching of whales on sea shores being observed in various regions around the world or similarly inexplicable cases of the deaths of sea mammals, fish and other hydrobionts in the White Sea, as well as the mass death of lemmings, etc., cannot be considered EC's. They are merely indicative of a troublesome ecological situation in the region.

Yet, the death of 4-20 million starfish *Asterias Rubens* on the Letnii Coast of the White Sea's Dvina Gulf in May 1990 is undoubtedly an EC since, as a result, practically the entire population of starfish was lost in the region.

However, it is little known that in 1979, in the same region, there was another catastrophe — the deaths of bottom-dwelling fish (flatfish, lancet fish, lumpfish and others). This catastrophe was no less in scale — the population of bottom-dwelling fish has not yet recovered. Yet, due to the former system of keeping all information on any kind of catastrophe secret, there was no press release nor an investigation into this matter (though it was prohibited to sell fish from the region).

Therefore we may assume as a fact that there were real EC's in the White Sea or at least phenomena close to them in scale.

We would like once again to analyse the probable role of toxicants in the development of these catastrophes, and to draw public attention to the problem of CW dumping in the White Sea. European countries closely follow the work of the Helsinki Commission established specifically for assessing the consequences of a probable leakage of CW dumped into the Baltic Sea. As for the Russian Federation Ministry of Nature Protection, it has not even started to work on evaluating the level of White Sea contamination with hazardous chemicals, though it is an internal sea of Russia. It is necessary to remember that consequences of a contaminants transfer from the White to the Barents Sea could affect fishery regions being used not only by Russia but also by Norway and other countries.

2.2. CHRONOLOGY OF THE 1990 EC IN THE WHITE SEA (LETNII COAST)

- May 4. On a water scoop lattice of the thermoelectric power plant (TEPP) in Severodvinsk, dead starfish were found, although only small ones (young). Simultaneously, dead young starfish appeared on the Letnii Coast (villages Sjuzma and Lopshenga).
- May 10 and 27. Vigorous storms threw ashore many starfish. Among them, none were alive. The starfish were of all ages, including very young that do not eat mussels. Beside starfish, there were dead mussels and crabs, and no fish.
- June 2. Strong storms threw out another round of starfish and mussels, all dead. The mussels were half-open.
- July 8. Another storm. Very few starfish were thrown ashore.

An examination done with an underwater apparatus showed that almost all starfish had been lost. In the observation area of the apparatus, 1-2 starfish per square meter were detected. The total number of dead and displaced starfish *Asterias Rubens* was estimated at 4-20 million.

That year (May — July) no fish came to Sjuzma for spawning. Local fishermen noted red stinking mucus on their nets.

A State Commission (06/10/90) noted the death of a girl who had been playing with starfish in Sjuzma. She was evacuated by helicopter to Arkhangelsk.

In September 1990 in the Onega Bay another lot of starfish was thrown ashore. They were all alive, however. The causes are unknown.

In 1979 in the same area, a mass death of bottom-dwelling fish was noted. The fish population has not yet recovered.

2.3. PROBABLE CAUSES OF THE CATASTROPHE

More than 10 hypotheses with different degrees of analysis were put forward to reveal the causes of the EC in the White Sea in 1990. We will discuss only two. Both of them were based on the findings of the State Commission having investigated this accident: "The death of starfish was caused by an unknown toxicant". However, due to the events of 1990-1991 no official conclusion of the Commission ever appeared.

It must be taken into account that an EC, similar to any other catastrophe, is the result of a sum of various factors which, taken separately, are not enough for such a drastic effect on nature. That is why, without denying the probable contributions of such causes as the general technogenic pollution of the White Sea, sea water dilution by fresh water during the flood in 1990, the loss of mussels thwart near the Letnii Coast and others, we choose a hypothesis based on a single-moment discharge of a drastic amount of a hazardous chemical.

In the Arkhangelsk Nature Protection Committee there are 2 volumes of documents collected by the State Commission. The analysis of their data allows us to suggest two types of toxicants: one of them - a strong reducing agent, the other - yperite (mustard gas).

2.3.1 *Reducing Agent — Hydrazine*

After receiving information on the EC the research vessel "Ivan Petrov" went to sea and performed a hydrochemical examination of the Dvina Gulf. The following anomalies were noted:

1. In the area of Severodvinsk - strong oppression of zoo plankton.
2. Elevated values of dehydrogenesis in bottom samples that indicate biological pollution of the ground. They found high concentrations of P, S, Fe, and elevated general mineralization. In all points dehydrogenesis activity - 0, in Sjuzma - 13, in Severodvinsk - 10.6.
3. At points 7, 9, 10 and 11 (highway Sjuzma — Severodvinsk), fermium ions were in reduced form. The relation of Fe^{2+} to the sum of ions [$Fe^{2+} + Fe^{3+}$] was 6-50% in all points, 85.68% in Sjuzma, 96.83% in Nenoksa, and 100% in Severodvinsk. This means that the concentration of reduced ions grows as one approaches Severodvinsk. All these data enable us to assume that the toxicant might be hydrazine, which is used for washing boilers at TEPP's. A hypothesis with a rocket fuel - dimethilhydrazine ("heptyl") - was absolutely rejected by the military. Hydrazine is highly toxic: its M.P.C. (maximum permissible concentration - *Ed.*) is 0.1 mg/l, the threshold concentration is 0.02 mg/l, lethal dose - 1-2 mg/l.

The Severodvinsk TEPP could be a large-scale toxicant source. A discharge of washing hydrazine could lead both to the deaths of oppressed sea organisms unable to migrate and to the reduction of metal ions. However, we should point out that the analysis of the Dvina Gulf's hydrodynamics makes this assumption rather vulnerable and a special verification by mathematical simulation methods is needed.

From this point of view an assumption on the use of heptyl rather than kerosene in so-called "flying objects" (FO, actually cruise missiles) seems true to life. In the period under discussion there were two incidents of FO's falling into the White Sea. On April 16, 1990, an FO fell into the sea approximately 100 km from the zone of the starfish catastrophe (fuel remainder was 166 kg). The second (June 6, 1990) fell very close to Nenoksa at a depth of 12 m, with a fuel remainder of 437 kg. A rough calculation for the value of intolerable hydrazine concentration (0.004 mg/l) shows that 400 kg of hydrazine form a 100 million cubic meter contamination zone, that is, it can be a layer of water 10 m deep, 20 m wide and 50 km long. Thus, the "hydrazine hypothesis" should not be rejected, though it needs a more accurate calculation.

2.3.2. *Toxicant — mustard gas*

An official report by the Arkhangelsk Fishery Complex indicates that, as a whole, they had 11 positive tests for yperite by the officially approved procedure of the Civil Defence Program (CDP). The tests were repeated and proved to be true. The Table 1 shows that after June 14, 1990 all tests were negative. A simple calculation for yperite, similar to that for hydrazine, on the basis of an intolerable concentration of 10^{-5} mg/l will give a contaminated layer of water of 10 m deep, 100 m wide and 10 km long. A more precise calculation will be given below.

TABLE 1. Analysis on the mustard gas, Arkhangelsk, 1990

The officially approved procedure of Civil Defense Program (CDP). Sensitivity 10^{-3} - 1^{-4} mg/l. (nt - not tested)

	Starfish	Herring	Mussels	Seaweed	Whitefish	Flounder	Navaga
May 23, Sjuzma	+	nt	nt	nt	nt	nt	nt
May 29, Sjuzma*	+	nt	nt	nt	nt	nt	nt
May 31, Sjuzma	nt	+	+	+	nt	nt	nt
June 4, Kozly	nt	+**	nt	nt	-	+**	nt
June 5, Marya	nt	+	nt	nt	nt	-	+
June 6, Petrominsk	nt	nt	nt	nt	nt	nt	+
June 7, Mudjug	nt	+***	nt	nt	nt	traces***	nt
June 14, Mudjug	ALL SAMPLES NEGATIVE RESULT						
June 24 , Lopshenga							
June 25, Lopshenga							
June 29, Mudjug and Marya							

*At the same time fish from the Barents Sea was tested, with negative result

**Analysis of these samples was repeated after one month's storage in a freezer. The result was the same.

***After slight roasting the probes were negative.

From May through June 1990, there were six strong storms with wind speeds of 10-14 m/sec in the Dvina Gulf, causing waves 2.4-3.3 m high [1]. This implies a strong mixing of coastal waters and motion along the shore toward Severodvinsk. Therefore, yperite and its hydrolysis products could be the reducing agent, the nature of which was discussed above.

3. Risk Assessment of a Single-Moment Leakage of Sunken CW

The history of the article on which this part of the report is based is rather remarkable. In 1994, the *Journal of Russian Chemical Society*, No. 2, was fully devoted to CW, and this paper was to be published in it, too. Yet, the paper ran so contrary to the opinion of the military authors on the danger of seal failure in sunken CW that the article was withdrawn, because this issue was financed by the military. The editorial board had to publish this paper in its next "civil" issue of the *Journal.*

The essence of our conflict is very simple: the military thinks that concerning the general contamination of the sea (in this case — the Baltic Sea), the role of CW discharges is negligible. This opinion could be correct if toxic agents (TA) were "common pollutants". However, they are not — TA are extremely hazardous toxic substances and their getting into water leads to absolutely different effects on living beings when compared with common pollutants (detergents, petrochemicals, municipal wastes, fertilisers, etc.).

We are not supporting alarmists, however, at the same time, we believe that the truth about the real danger from sunken CW must be nether hidden nor underestimated.

The question about the possibility of a single-moment seal failure of a piece of chemical ammunition or a container of TA either greatly irritates military specialists or is not answered at all. We can think of a lot of reasons for such event: a strike with an anchor, bottom trawl or any other object; a cracked container may break under depth pressure. Many bombs have some floating ability which enables them to move downstream and reach shores or shallow water where they can be broken by tidal or storm waves, and so on and so forth.

It is such considerations that incline us to perform at least an approximate calculation of the probable results of such leakage of sunken CW.

As ecological factors we used:

- Lethal dose, LD_{100}, (Zone A);
- Intolerable concentration, C_{int}, (M.P.C. for an industrial zone) (Zone B);
- "Initial" (threshold) concentration, C_{ini} (M.P.C. for residential areas) (Zone C);
- "Under threshold" concentration, C_{und} (Zone D).

The relations of concentrations in these zones are: 100:1:0.1:0.01

As "initial" we assume the concentration that causes an alarming reaction with hydrobionts, while in Zone D the danger due to the cumulative effects remains, the action being prolonged.

When describing ecological effects on sea inhabitants, one should take into account their mobility. It is obvious that fish will leave both Zone A and Zone B, and probably Zone D. Immobile populations will be extinguished in Zone A, partially destroyed in Zone B and oppressed in Zone C.

We do not consider the toxicity of hydrolysis products, thus underestimating the danger. For example, in the hydrolysis of sarin, several compounds are formed, and among them only one can be considered non-toxic. In the case of yperite, hydrolysis products are also highly toxic, and, as for lewisite, oxide produced in hydrolysis is equally dangerous.

Here lies the departure between the military business and ecology. When military experts say that yperite cannot be transferred by trophic chains and accumulated in organisms [3], they proceed from the fact that yperite molecules disappear upon penetrating an organism, so they give no more thought to it. Yet, the active group of yperite - SCH_2CH_2Cl remains, and the accumulation of organochlorine substances in lipids is well-known.

Unfortunately, at present I cannot make an accurate calculation to get a map of TA distribution, though it is possible to give well-grounded approximate values based only on the known data on TA hydrolysis.

We use a simple algorithm:

1. The volume of the water layer with an "intolerable" concentration is calculated from the TA amount. Data: $C_{int} = 10^{-2}$ mg/m^3, $P_{st} = 100$ kg.
2. The time of formation of the C_{int} layer is determined from stream velocity. Data: V_{str}=1-5 m/min.
3. The loss of TA due to hydrolyses during this time is found from the constant of hydrolysis rate of certain TA to the level of C_{in}. Data: k=0.1 day^{-1} at 10° C for soman, k=0.014 min^{-1}at 10° C for yperite.
4. Correction for solubility is made. Data: yperite 0.06% at 10° C (600 mg/l), lewisite 0.005% at 25° C (0.5 kg/m^3).

Let as discuss the behaviour of three main types of TA during leakage: organophosphorous toxic agents (OPTA), yperite (mustard gas) and arsenic-containing TA.

3.1. SARIN

A simple calculation for sarin shows that single-moment leakage of one CW bomb with 100 kg of TA, in the conditions indicated above (but with k = 0.308 day^{-1}), will lead to intolerable contamination of a water volume of 10^{10} m^3. That means a water layer 100 m high (typical depths of the Baltic Sea), 1000 m wide and 100 km long. Neglecting internal diffusion (layer formation) and considering only the drift of TA with the stream, we arrive at the following result: at v = 1 m/min it takes 40 days to form a layer 100 km in length and, respectively, at v = 5 m/min - a bit more than 12 days. However, hydrolysis will lead to reduction of TA concentration and in approximately 15 days (99% conversion) only 1% (1 kg) of the original 100 kg of TA will be left. This means that the concentration within the layer will reduce by two orders in relation to the intolerable concentration. In fact, the TA concentration will decrease less significantly. It results from two factors:

1. The hydrolysis rate was determined in a closed system characterised by autocatalysis which does not exist in open systems.

2. Real temperature at sea depths can be assumed as 10° C (in reality only 5° C), this being 15° C lower than the temperature of the hydrolysis rate constant used for the calculation. Assuming that the hydrolysis constant has a normal temperature dependence (E_a = 50-80 kJ/mol), we will have k=0.1 day^{-1} or less (complete hydrolysis of soman at 18 °C takes more than 2.5 months).

If we take into account these two factors, it is possible to assume that, even a half-month after leaking in a huge body of water, TA concentration will be at least "initial".

If stream velocity is five times lower, an "intolerable" concentration will be maintained for more than two weeks in a layer 100 m high, 500 m wide and over 40 km long.

3.2. YPERITE

If we carry out the calculation for yperite (mustard gas) the result will be similar. Yperite's solubility in water at 0° C is 0.03% and 0.08% at 20° C. Assuming the solubility value at 10° C is 0.06%, we arrive at the following: the concentration of saturated solution will be 600 mg/l, that is, it exceeds lethal doses by thousands of times. For example, irreversible damage to the eyes takes place at concentrations as low as 0.001 mg/l. This permits us to assume as "intolerable" concentration the level of 10^{-5} mg/l - similar to other TA. The difference from OPTA is that yperite can maintain a stationary concentration (600 mg/l) in a flowing water layer for a long time, thus increasing the danger of yperite. However, its hydrolysis rate is 60 times higher than that of sarin. This circumstance decreases the volume of the toxicated layer by one or even two orders, though the hydrolysis products of yperite are much more toxic than those of sarin. In addition, yperite is characterised by a high degree of adsorption and penetration into various materials, thus prolonging the existence of the toxic zone.

In summary, we can say that the leakage of yperite ammunition leads to the toxification of a lesser water volume, though for a longer period of time.

3.3. ARSENIC-CONTAINING TOXIC AGENTS

The third group of TA is represented by arsenic derivatives: lewisite and three irritants - diphenylchloroarsine, diphenylcianarsine and adamsite.

Adamsite has a low solubility in water and is very stable to hydrolysis. Dihydrophenarsasine oxide, a product of adamsite's hydrolysis, is no less active an irritant than the original adamsite. Adamsite, if it gets into water, will consistently poison the underwater vicinity of the buried TA. This is connected with the fact that the established limit of the notion "insoluble in water" is 1:>10,000, but if we assume its solubility even as 1:100,000 this means a concentration of 0.01 g/l. This value is only one order lower than an "intolerable" concentration, and it will inevitably cause toxification of biota.

Lewisite's behaviour at leakage could be reminiscent of yperite: it also has low solubility in water (0.05% at 20° C) and is considerably heavier than water (density 1.88 g/ml at 20° C). Yet, it is much more reactive than yperite and it is easily hydrolysed, thus the affected zone becomes even smaller than that of yperite. However, a specific feature of lewisite is its hydrolysis resulting in an oxide of the same toxicity. This oxide is a solid substance practically insoluble in water. The example of adamsite has just shown us the relativity of the notion of insolubility.

The total aggregate of the physical, chemical and toxicological properties of arsenic-containing TA leads to the formation of stable poisoned areas with solid hydrolysis products (up to inorganic arsenic compounds) in locations of TA burials. It is evident that life is impossible in these zones.

Considering all three groups of TA after leakage allows us to draw the following conclusions:

1. Water-soluble OPTA cause the toxification of huge water volumes affecting all organisms. In the range of OPTA activity on large areas (zone A and zone B) living beings will be annihilated.
2. Low-soluble yperite (mustard gas) intoxicates a smaller water volume, but its toxicity remains for a longer period. Also, the absorption of yperite by plants and phytoplanktone allows it (in various forms of cell alkylates) to penetrate food chains, which is typical of organochlorine compounds. This is especially dangerous due to the strong carcinogenic and mutagenic characteristics of yperite. Both in the first and second cases, the cumulative effects should be considered.
3. Arsenic-containing TA form an even smaller volume of toxicated water, but they create a stationary poisoned spot where life is impossible.

According to the calculations done by military experts in [3], the maximum rates of leakage of CW sunk near Bornholm Island will be 40 tonnes/year, that is nearly 100 kg/day, 60 years after burying (1947), and 256 tonnes/year, or nearly 700 kg/da, - 125 years after (second leakage maximum). During the second maximum, more than 6,000 tonnes of yperite (85% of the sunk amount) will be discharged into the water. These processes will go on for more than 200 years. This is our “present” to future generations.

4. Conclusion

The given analysis clearly shows our insufficiency of knowledge not only about points of CW sinking, but also about the influence of CW leakage on the environment. At the same time, the great threat of chemical ammunition leakage to all sea organisms and, therefore, to man, is obvious.

It seems absolutely necessary to:

1. collect data on CW dumpings;
2. carry out laboratory-scale studies of the kinetics of TA hydrolysis in real salt solutions;
3. perform strict toxicological studies on the effect of threshold ("initial" and "intolerable") concentrations of TA and their hydrolysis and transformation products on sea organisms. In addition, it is necessary to study the influence of the prolonged action of "under-threshhold" concentrations on the biota's survival. Within biological investigations, the rate of biota reproduction upon being affected by various TA must be assessed;
4. organise several expeditions to the points of CW sinking for collecting data on the state of both the sunken ammunition and the surrounding biota.

References

1. *Priroda* **6**, 1991.
2. *Journal of Russian Chemical Society* **2**, 1994, p. 80.
3. *Journal of Russian Chemical Society* **2**, 1994, p. 114.

Index

—O—

—P—

—R—

—S—

—T—

—U—